# Medical Equipment Management

# Series in Medical Physics and Biomedical Engineering

Series Editors: John G Webster, E Russell Ritenour, Slavik Tabakov,
Kwan-Hoong Ng, and Alisa Walz-Flannigan

Series in Medical Physics and Biomedical Engineering

# Medical Equipment Management

**Keith Willson**
*Imperial College London, UK*

**Keith Ison**
*Guy's and St Thomas' Hospital, London, UK*

**Slavik Tabakov**
*Kings College London, Strand, UK*

**CRC Press**
Taylor & Francis Group
Boca Raton London New York

CRC Press is an imprint of the
Taylor & Francis Group, an **informa** business
A TAYLOR & FRANCIS BOOK

CRC Press
Taylor & Francis Group
6000 Broken Sound Parkway NW, Suite 300
Boca Raton, FL 33487-2742

Printed on acid-free paper
Version Date: 20130905

International Standard Book Number-13: 978-1-4200-9958-4 (Hardback)

### Library of Congress Cataloging-in-Publication Data

Willson, Keith, author.
  Medical equipment management / Keith Willson, Keith Ison, and Slavik Tabakov.
    p. ; cm. -- (Series in medical physics and biomedical engineering)
  Includes bibliographical references and index.
  ISBN 978-1-4200-9958-4 (hardcover : alk. paper)
  I. Ison, Keith, author. II. Tabakov, Slavik, author. III. Title. IV. Series: Series in medical physics and biomedical engineering.
    [DNLM: 1. Equipment and Supplies. 2. Maintenance. 3. Materials Management, Hospital. 4. Technology Assessment, Biomedical.  W 26]

  R857.M3
  610.28'4--dc23                                                          2013032824

**Visit the Taylor & Francis Web site at**
**http://www.taylorandfrancis.com**

**and the CRC Press Web site at**
**http://www.crcpress.com**

# Contents

# Series Preface

The *Series in Medical Physics and Biomedical Engineering* describes the applications of physical sciences, engineering and mathematics in medicine and clinical research.

The series seeks (but is not restricted to) publications in the following topics:

- Artificial organs
- Assistive technology
- Bioinformatics
- Bioinstrumentation
- Biomaterials
- Biomechanics
- Biomedical engineering
- Clinical engineering
- Imaging
- Implants
- Medical computing and mathematics
- Medical/surgical devices
- Patient monitoring
- Physiological measurement
- Prosthetics
- Radiation protection, health physics, and dosimetry
- Regulatory issues
- Rehabilitation engineering
- Sports medicine
- Systems physiology
- Telemedicine
- Tissue engineering
- Treatment

The *Series in Medical Physics and Biomedical Engineering* is an international series that meets the need for up-to-date texts in this rapidly developing field. Books in the series range in level from introductory graduate textbooks and practical handbooks to more advanced expositions of current research.

The *Series in Medical Physics and Biomedical Engineering* is the official book series of the International Organization for Medical Physics (IOMP).

## The International Organization for Medical Physics

The IOMP represents over 18,000 medical physicists worldwide and has a membership of 80 national and 6 regional organisations, together with a number of corporate members. Individual medical physicists of all national member organisations are also automatically members.

The mission of the IOMP is to advance medical physics practice worldwide by disseminating scientific and technical information, fostering the educational and professional development of medical physics and promoting the highest quality medical physics services for patients.

A World Congress on Medical Physics and Biomedical Engineering is held every three years in cooperation with the International Federation for Medical and Biological Engineering (IFMBE) and the International Union for Physics and Engineering Sciences in Medicine (IUPESM). A regionally based international conference, the International Congress of Medical Physics (ICMP), is held between world congresses. IOMP also sponsors international conferences, workshops and courses.

The IOMP has several programmes to assist medical physicists in developing countries. The joint IOMP Library Programme supports 75 active libraries in 43 developing countries, and the Used Equipment Programme coordinates equipment donations. The Travel Assistance Programme provides a limited number of grants to enable physicists to attend the world congresses.

IOMP co-sponsors the *Journal of Applied Clinical Medical Physics*. It publishes, twice a year, an electronic bulletin, *Medical Physics World* and an electronic Journal, *Medical Physics International*. It also publishes e-Zine, an electronic newsletter, about six times a year. IOMP has an agreement with Taylor & Francis Group for the publication of the *Medical Physics and Biomedical Engineering* series of textbooks. IOMP members receive a discount.

IOMP collaborates with international organisations, such as the World Health Organization (WHO), the International Atomic Energy Agency (IAEA) and other international professional bodies, such as the International Radiation Protection Association (IRPA) and the International Commission on Radiological Protection (ICRP), to promote the development of medical physics and the safe use of radiation and medical devices.

Guidance on education, training and professional development of medical physicists is issued by IOMP, which collaborates with other professional organisations in the development of a certification system for medical physicists that can be implemented on a global basis.

The IOMP website (www.iomp.org) contains information on all the activities of the IOMP, policy statements 1 and 2 and the 'IOMP: Review and Way Forward', which outlines all the activities of the IOMP and plans for the future.

# *Preface*

Any organisation delivering healthcare with modern medical technology relies to a greater or lesser extent on medical devices to benefit patients. As technology in clinical care becomes more complex, so do demands on those managing the equipment which delivers it. Clinical professionals rely on equipment to provide what it promises and not let them down. They and the clinical engineers, technologists and managers who support them need to know how to keep patients safe and equipment working in the clinical environment.

This book is based on topics included in an MSc course in clinical engineering and medical physics offered at King's College London since 2005. The material has since been expanded and developed considerably. Although based on a postgraduate course, the text does not contain any difficult technical concepts and is structured to be accessible to all healthcare professionals and managers. It does not remove the need for a reader to seek expert advice on specific legal or technical matters but rather seeks to explain underlying principles and requirements to raise awareness of what needs to be done and what questions to ask. It therefore provides practical advice and refers the reader to appropriate legislation and guidelines without attempting detailed or complex interpretations. We consider it to be of value to clinical engineers and technologists, clinicians, managers, laboratory workers, procurement and estates staff and established professionals from all healthcare disciplines. The material is relevant to professional training schemes for clinical engineers and scientists and will assist students and trainees undertaking MSc, BSc, MEng, BEng or diploma courses in clinical engineering, clinical technology, medical physics and related subjects. We hope that you find it useful and enjoyable.

# Acknowledgements

We are grateful to our colleagues at Guy's and St Thomas' Hospital, King's College Hospital, King's College London and Imperial the Charing Cross Hospital, who have contributed to developing the contents of this book. These include Bruce Bartup, Peter Cook, Prabodh Patel, Emmanuel Akinluyi, Mayur Patel, Rashid Brora, Wakulira Arafat, Nick Abraham and Mike Sweeney. We would like to thank Graham Jackaman and Neville Fowler for detailed advice on decontamination issues and Katrina Cooney and Prabodh Patel for their review of manuscript drafts. We are particularly grateful to Peter Cook for his extensive and detailed comments. We would also like to thank Nigel Pearson for his time and skill in photographing medical equipment and Joe Cook and Avtar Verdee for access to imaging equipment. We also appreciate the support we have been given by Guy's and St Thomas' NHS Foundation Trust, Imperial College Healthcare NHS Trust, King's College London and King's College Hospital NHS Foundation Trust.

We are grateful for encouragement and direction we have received from the editorial and production team at Taylor & Francis Group, especially to John Navas for his support for this project over its extended lifetime and for invaluable input from Rachel Holt and Francesca McGowan.

# *Authors*

**Keith Willson** earned his BSc in electrical and electronic engineering from City University (London) in 1972 and his MSc in medical electronics and physics from St Bartholomew's Medical School (London) in 1974. His career has been dedicated to medical device technology, working in hospital, university and manufacturing environments. He headed the Clinical Engineering Department at Royal Brompton Hospital from 1996 to 2007 with responsibility for hospital-wide equipment management and the provision of specialised devices for a wide range of research projects. He is currently employed as a principal research fellow at the UK National Heart and Lung Institute, providing an engineering input into research on therapies for periodic breathing, cardiac pacemaker optimisation and novel developments in ultrasound image analysis. He teaches healthcare technology management on MSc courses at City University and at King's College London.

**Keith Ison** graduated in physics and materials science from the University of Cambridge before training as a medical physicist in Hull. After completing his PhD in biomaterials at Bath University, he worked as a postdoctoral researcher on cochlear implants and then in a variety of NHS scientific roles in clinical measurement, audiology and management while studying for an MBA with the Open University. He spent ten years at King's College Hospital, including three years as a divisional general manager, before becoming head of medical physics at Guy's and St Thomas' Hospitals in 2001. He is responsible for a wide range of medical physics and clinical engineering activities and is an honorary senior lecturer at King's College London, organising the Equipment Management module on the MSc in Medical Engineering and Physics. He is a past president of the Institute of Physics and Engineering in Medicine and is active in the development of education and training of scientists in healthcare. He was awarded an OBE for leadership development for engineers and scientists in healthcare in 2012.

**Slavik Tabakov** is currently programme director of MSc courses in medical engineering and physics at King's College London and consultant medical physicist in the Department of Medical Engineering and Physics, King's College Hospital. He became president-elect of the International Organization for Medical Physics (IOMP) in 2012 and is editor of the IOMP journal *Medical Physics International*. He has had a long involvement with the education and training of physicists and engineers and has chaired the education and training committees of IOMP, International Federation for Medical and Biological Engineering (IFMBE) and the International Union for Physics and Engineering Sciences in Medicine (IUPESM). He is a leading developer of the concept of international e-learning in medical physics, both of the first educational image database and e-books which are currently used in over 70 countries and also of the first e-encyclopaedia of medical physics with its related multilingual dictionary of terms produced in 29 languages. He was awarded the IOMP Harold Johns Medal for Excellence in Teaching and International Education Leadership in 2006. He is an expert on various IAEA projects and advises on medical physics education in 15 countries.

# 1

## Introduction

### 1.1 Scope of Medical Equipment Management

Medical equipment is an indispensable part of healthcare. Technological advances continue to increase its capability and safety and improve patient care. These advances are also having wider consequences. The total cost of most organisations' medical equipment is increasing quickly, as are the clinical and financial impacts of device failure or operator error. Concerns over return on investment and value for money have affected attitudes towards equipment procurement and expanding government regulation, with clinical and safety targets and reduced public tolerance of adverse incidents leading to a greater awareness of equipment-associated risks amongst clinical staff, managers, patients and the public. Government policies, professional bodies, risk managers and insurers all demand that medical devices are actively managed throughout their lifetime, from purchase to disposal. One of our intentions in writing this book is to raise awareness of the importance of medical equipment management and encourage the active and informed participation of all healthcare staff in improving the way equipment is purchased, used and cared for.

Medical equipment management has three essential aims: To ensure medical equipment in the clinical environment is appropriate to the needs of the clinical service, that it functions effectively and safely and that it represents value for money. Its remit extends far beyond equipment repair and maintenance to include clinical governance and risk and asset management. Effective medical equipment management is always a collaborative process between different groups. Successful acquisition of medical equipment, for example, involves clinical, technical and financial evaluations carried out by a wide range of qualified specialists, undertaken with regard to financial and corporate governance and in compliance with procurement regulations.

The life of an item of medical equipment starts with demonstrating a need for it, finding funding, writing specifications, tendering and selecting the best option from those models on offer. When an item is delivered, there should be an acceptance procedure to check it works correctly and meets safety standards. Users need to be trained and equipment has to be maintained and supported. Attendant risks have to be managed, including clinical risks from operator misuse and malfunction and safety risks to patients and staff. Research projects require specialised support to identify and address new risks. Adverse incidents and near misses need to be investigated promptly and corrective and preventive action taken. Even at the end of its life, equipment cannot simply be thrown away but must be condemned and disposed of according to good corporate governance principles whilst addressing environmental and health and safety regulations.

Each activity requires input from qualified and trained staff and the whole medical equipment management system must be directed and supported by the healthcare organisation which commissions it. Devoting resources to equipment may not be seen as high priority by an underfunded organisation, especially where it seems to compete with clinical priorities, yet healthcare organisations rely increasingly on technology to deliver clinical care and cannot ignore it. They are obliged to promote good medical equipment management and comply with various important requirements which, as we shall see, include legislation, standards and best practice. Medical equipment management is not, though, just a matter of responding to compulsion. There are positive operational and financial benefits to be gained from it. An example is standardising equipment, which brings financial and operational benefits from economies of scale in training and maintenance and reduces the risk of errors made by clinical and technical staff working with unfamiliar equipment.

## 1.2  Who Should Read This Book?

In a climate increasingly sensitive to the way medical devices are used, all those working with or supporting medical equipment need to understand their responsibilities and the financial and organisational risks and implications of what they do. Clinical professionals, engineers, technologists, managers and clinical directors all need to know how best to use and care for medical devices in the interests of patients and their own organisation.

Every hospital and health facility has at least one person responsible for managing medical equipment and many clinical professionals carry out elements of this role in departments and management units. These individuals work with specialist engineers, technologists and scientists to deliver effective medical equipment management and will benefit from a greater awareness of underlying issues, as will ward and corporate nurses, clinical supervisors and departmental managers who are responsible for the safe use and care of equipment.

Many graduate- and postgraduate-level courses are being offered which include medical equipment management as a course module or core content, including those delivered to individuals working for independent providers and suppliers. In addition, there is a growth of self-learning and professional development in this discipline as the scope of what it covers and rate of technological development increase. This includes engineers, scientists and technologists in training and those working in hospitals, independent healthcare and commercial organisations at all stages of their careers including education, training and professional practice. Much of the material is relevant to professional training schemes for clinical engineers and scientists and so will assist students and trainees undertaking MSc, BSc, MEng, BEng or diploma courses in clinical engineering, clinical technology, medical physics and similar subjects.

We aim therefore in this book to provide a text that will be of value to all those engaged in medical equipment management. This book will appeal to a wide range of professionals, not only to those who are technical and clinical but also to healthcare managers at all levels and those responsible for, or overseeing, equipment procurement and management. These include clinical general managers and administrators, together with those in supplies, estates and finance departments. It has something to offer to all professionals who want to broaden their understanding or expand their professional profile in this area.

Many modern items of medical equipment incorporate embedded control and analysis software, and link to data storage and review systems. These devices are increasingly being connected to IT networks, for example, local point of care diagnostic testing devices such as blood gas analysers sending results directly into the electronic patient record (EPR), physiological data storage and analysis systems feeding into central data repositories via IT links and image acquisition and storage systems supporting remote clinical reporting. Therapy is not excluded, with radiotherapy installations programme and control radiation delivery across IT networks, intensive care monitoring being carried out over extended areas and surgeons controlling robotic manipulators over communication links to remote sites. All of these developments are changing the way diagnosis and therapy can be delivered and are having a major impact on healthcare technology and its management.

## 1.3 Approach and Content of This Book

National and international laws and standards in medical devices are complex and relate to specialised areas of design, performance and management. Those with medical equipment management responsibilities must understand what is important and seek guidance on how to translate regulation and guidance into everyday practice.

This book therefore goes beyond principles and requirements to look at the practice and politics of equipment management. It is structured in sections so that individuals can access it in the way that best suits them. For example, healthcare managers can gain a good overview of equipment management processes from the relevant chapters without needing to read those sections containing technical and process details aimed at engineers and students. It is organised into four broad elements:

- *Introductory chapters* are intended for general readers, including clinical staff, healthcare managers and others looking to get an initial overview.
- *Specialised chapters* are devoted to specific technical and managerial topics but contain introductions and summaries to guide the reader to areas of particular interest.
- *Appendices.*
- *Examples* provide practical illustrations of how to realise various aspects of medical equipment management and contain material of value to all readership groups.

The introductory chapters are Chapters 2 and 3. Chapter 2 provides an overview of the medical equipment life cycle in order to signpost the contents of this book to the reader. We trace three different items of equipment through their life cycle from procurement to disposal, addressing issues as they arise and providing pointers to the chapters and sections where relevant further material is presented. This provides a broad-based introduction, accessible to non-specialists and those new to the field, which puts equipment management into a healthcare context. Chapter 3 then examines the implications of governance, quality, safety and risk concepts for medical equipment management and for clinical engineering. We put these into the context of legislation and organisational policies to show how this leads to the equipment management processes described in Chapter 2. We introduce the benefits of a structured approach for the patient and the organisation and highlight some of the pitfalls and consequences of not getting it right. This second introductory chapter will also help to guide managers and those carrying responsibility for equipment management to chapters in the book they can consult for further help.

The remaining chapters are devoted to various equipment management topics and contain more detailed expositions of the relevant principles, covering all major aspects of medical equipment management. It is intended that each chapter can be accessed and understood independently, with minimal reliance on other chapters. Chapter 4 considers the context around medical equipment management in a healthcare context and the systems developed to support it, including internal governance processes, and considers the balance between internal and external provision of medical equipment management services. Chapter 5 deals with equipment acquisition, including specification and tendering, and explores how to allocate resources across competing equipment needs. Chapter 6 summarises the procurement process and highlights how medical device specification and evaluation can be carried out in this context. Chapter 7 looks at the need to train clinical and technical staff and patients in the use and support of equipment and considers how the complex logistics of providing this training can be implemented. Chapters 8 and 9 discuss equipment maintenance. In Chapter 8, we explore how risk assessment and cost–benefit analysis can identify the best approach to specifying the frequency and nature of maintenance and in Chapter 9, we look at factors to be taken into account when deciding whether to perform maintenance in-house or outsource it. We consider the advantages and disadvantages of the various types of service contract and managed services available and suggest how these can be monitored effectively.

Adverse incidents and near misses arising from medical device use need to be reported and investigated. Their causes should be identified and associated issues analysed so that action can be taken to prevent any recurrence or other associated problems. Chapter 10 introduces this topic and gives practical advice on risk assessment, investigation techniques and preventive actions. Chapter 11 looks at ways to manage medical equipment and associated risks in research and innovation including new and novel applications, showing how an organisation can reduce the liability associated with such equipment use. Risks do not end once equipment has reached the end of its working life, and Chapter 12 introduces the practical, regulatory and governance issues associated with responsible equipment disposal.

The final two chapters and appendices address specialised knowledge in topic areas important to those engaged in providing a high-quality medical equipment management service. Chapter 13 introduces a wide range of regulatory bodies, standards organisations, professional associations and other sources of advice and assistance that enable the medical equipment management professional to practise safely and effectively throughout their career. In Chapter 14, we look at the performance measurement of medical equipment management processes, describe the kinds of external scrutiny equipment management services are subject to and consider practical aspects of performance indicators and benchmarking. Appendix A sketches out how a practical medical device service might work and Appendix B looks in detail at electrical safety, an area relevant to many types of medical equipment.

Although many examples in this book deal with standards and practice in the United Kingdom National Health Service (NHS) and Europe, principles of medical equipment management apply worldwide. We have aimed therefore to provide a text of value to a global readership by identifying principles alongside specifics and encourage the reader to apply these principles in their own context. Medical equipment management processes and topics, including quality, risk and safety, are universal but subject to local legislation, guidelines and emphasis. In some cases, local standards apply internationally: For example, equipment produced in the United States will need to meet European regulations to be sold there and vice versa, making European and US regulations relevant to equipment manufacture and performance in both countries. Similarities exist too between medical device regulatory approaches, for example, between China, the European Union and the United States. Indeed, some nations rely in part on regulatory mechanisms used by other countries, and the importation of medical devices into a number of developing countries is dependent on equipment meeting US requirements. Increasing globalisation of trade in medical devices is driving harmonisation of standards and a convergence of national approaches to medical equipment management.

## 1.4  Clarifications

Throughout this book, we have used the terms *medical devices* and *medical equipment* widely. Chapter 2 tries to define the scope of these terms, as the former includes the latter along with other items such as consumables and medical software. We have chosen not to call this book *medical device management* as much of what we have written about is relevant to medical equipment, although we do consider issues of relevance to medical device items as they arise. We also talk about *organisation* and *healthcare organisation* rather than *hospital*, as much of what we say is as relevant to managing equipment in a local clinic as it is to a teaching hospital.

We recognise that medical equipment management has a lot in common with equipment management in general, and asset management principles are described and explored further in documents from professional and standards bodies [1–2]. It does however have its own distinctive and special features, and we have tried to bring these out in this text.

## 1.5  Values and Value

All health systems in the developed world are running into problems to do with resource limitations. Demand for healthcare is outstripping the ability of nations to pay for it. Difficult decisions are having to be made on the basis of relative benefit and cost, judgements which rely either explicitly or implicitly on ethical judgements and values frameworks, and which are worked out through political processes. Individuals working with medical equipment must not lose sight of the reason why health technology is in use in the first place, and should be aware of not only the financial impact of what they do but also of the clinical value to patients that is added by every medical equipment management activity they undertake.

## References

1. Public Access Statement 55-1. *Asset Management Part 1 – Specification for the optimised management of physical assets*. British Standards Institute, London, 2008.
2. ISO 55000 International Standard for Asset Management. International Organisation for Standardisation, Geneva.  http://www.assetmanagementstandards.com/ (accessed on September 09, 2013).

# 2

## Medical Equipment and Its Life Cycle

### 2.1 Introduction

Medical equipment takes time and effort to manage. In this chapter, we introduce its variety and complexity by following three different items of equipment through their life cycle, starting from the initial awareness that they are needed to final decommissioning and disposal perhaps a decade or more later. These three items are a large batch of syringe drivers, an ultrasound scanner and an x-ray facility. We have chosen these devices to illustrate the variety of effort and resource required to take different types of equipment through their operating life. Along this journey, we introduce key equipment management issues and point to other parts of this book containing more detailed material. This broad-based introduction is intended to be accessible to specialists and non-specialists alike, whatever their background or discipline.

Looking at the broad sweep of medical equipment management illustrated in Figure 2.1, we start by considering how to demonstrate the need for new equipment and find funding for it. We describe formal procurement processes, how these are supported by clear specifications and tendering and how different models can be compared to choose the best one. We show that acceptance and commissioning require thought, planning and preparation, whether dealing with single or multiple devices or complex installations. User training and support is vital to keep equipment operating effectively, and we show that effort in these areas is as important as good maintenance in managing risk. We introduce the differences between contract and in-house maintenance and show that, even at the end of its useful life, equipment must be condemned and disposed of with due regard for organisational governance and external regulation.

We start by defining the term *medical equipment* and then outline each of the major equipment management processes and the people, skills and timescales involved. Each of these processes and the management techniques needed to address them are covered in more detail in later chapters. We finally introduce the concept of *clinical engineering* and the role of the clinical engineer in managing medical equipment and in helping other healthcare professionals with this task. Healthcare organisations should support

**FIGURE 2.1**
Overview of medical equipment management.

good medical equipment management to meet regulations, standards and best practice, and we describe some benefits of good practice for patients and the organisation and also consequences of its poor delivery.

## 2.2 What Is Medical Equipment?

We quote in the following paragraph definition of a medical device set out in the European Medical Devices Directive 2007 [1] which is similar to that of the World Health Organization (Chapter 13). Detailed definitions vary between countries and these differences have legal implications for manufacturers and for equipment management practices in the states to which they apply.

*Definition of a medical device*:

> Any instrument, apparatus, appliance, software, material or other article, whether used alone or in combination, together with any accessories, including the software intended by its manufacturer for its proper application intended by its manufacturer to be used specifically for diagnostic and/ or therapeutic purposes and necessary for its proper application, intended by the manufacturer to be used for human beings for the purpose of
>
> - Diagnosis, prevention, monitoring, treatment or alleviation of disease,
> - Diagnosis, monitoring, treatment or alleviation of or compensation for an injury or handicap,
> - Investigation, replacement or modification of the anatomy or of a physiological process,
> - Control of conception,
>
> and which does not achieve its principal intended action in or on the human body by pharmacological, immunological or metabolic means, but which may be assisted in its function by such means.

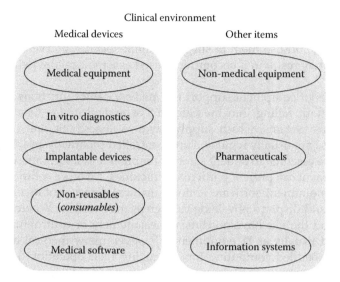

Clinical environment

Medical devices  Other items

Medical equipment  Non-medical equipment

In vitro diagnostics

Implantable devices  Pharmaceuticals

Non-reusables (*consumables*)

Medical software  Information systems

**FIGURE 2.2**
How medical equipment relates to other items in the clinical environment.

Despite this wide-ranging definition, there is still a scope for ambiguity as to whether some kinds of device or their associated equipment should be classified as *medical devices*. Figure 2.2 summarises how types of device in the medical environment can be classified. Looking at intended function helps to identify some of the ambiguities, and examples are as follows:

- *Prevention* includes inoculation devices but not items solely to prevent injury or staff infection, which are better classed as, *personal protective equipment* (PPE). Surgical gloves are both medical devices and PPE and can be classified as both in Europe.

- *Diagnosis and monitoring* includes in vitro laboratory and point of care or near-patient testing devices. Some devices sold for independent self-monitoring, such as blood pressure monitors aimed at runners, are not sold as *medical devices* but *for indication only*, yet may still find their way into the medical environment.

- *Devices with a pharmacological component*, for example, drug-eluting stents, are usually easier to develop and market than a new pharmaceutical. More devices of this type are being developed, and their classification can provide a challenge to national regulators.

- *Replacement or modification* includes both passive and active implants, such as pumps designed to release drugs over extended timescales. Passive filler material for plastic surgery may be classed as a medical device depending on its final purpose. If used for cosmetic purposes its use may be controlled differently or not at all.

- *Software for managing patient records* is not a medical device under the definition above but if a module is added to it to extract clinical details and use them to suggest patient-specific interventions, then that may result in it becoming classed as a medical device.

Other types of equipment support the medical device functions highlighted above without falling into the category of *medical device*. One example is medical gas systems, which supply gases to anaesthetic and other equipment. These are usually regulated and controlled separately and, like medical devices, are essential to patient safety. Computer and information technology (IT) equipment can present particular difficulties. An office computer running patient appointment software is not a medical device, but a computer that processes and displays physiological signals may well be. Where computers, printers and the like have no clinical measurement or control function, they are classed usually as IT equipment. However, many medical devices now contain embedded computing for controlling their operation or processing physiological signals, and many are directly connected to computers or implemented as software on them. Where IT equipment is essential to a medical device function, it is classified either as a medical device or accessory. Software is considered to be a medical device if it performs or enables a medical device function. This includes *stand-alone software* used, for example, to process diagnostic images on an independent imaging workstation [2]. Medical equipment can be linked to an IT network, for example, to store images or patient measurements on a central server. In these circumstances, the IT network is not a medical device but needs to be managed appropriately to ensure the correct operation of medical equipment and minimise associated risks.

A healthcare organisation needs to be clear what is included in the legal definition of a medical device, in order to manage its equipment appropriately. Medical laboratory equipment, for example, may be overlooked because a clinical engineering department is concerned mostly with maintenance and support for front-line patient devices. Conversely, if a clinical engineering service is part of a wider estates and facilities function, items used in the patient area that are not medical devices may be perceived as such, and clinical engineering services might apply inappropriate controls to facilities such as air conditioning, electrical power supplies and medical gas pipelines which have their own specific regulations. Guidance is available from national regulators (Chapter 13).

## 2.3 Equipment Management Processes

We now describe the various medical equipment management processes in more detail. Figure 2.3 shows how these fit into the equipment life cycle.

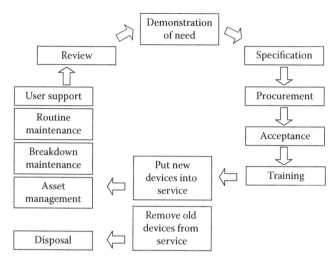

**FIGURE 2.3**
Detailed medical equipment life cycle.

### 2.3.1 Establishing Need

Before equipment is acquired, the need for it has to be demonstrated (Chapter 5). This is easier to do where items are a straightforward replacement and more difficult when they are additional. A replacement request often comes from end users who think equipment is outdated or unreliable or may come from a manufacturer or clinical engineer facing difficulties in supporting older equipment. Devices entirely new to the organisation might be needed to set up a new clinical service or to carry out research. These are usually requested by clinicians but need support from senior management where they are part of a strategic clinical initiative. Under leasing or managed equipment arrangements, equipment renewal is assessed at agreed intervals.

An end user may be very clear and vocal about the need for an item but is unlikely to be the only person in their organisation asking for new equipment. The cost and priority of each item must be judged against overall clinical need, available budget and the balance between maintaining existing services and the need to innovate. In the interests of economy, alternatives to direct purchase may be considered such as upgrading equipment or redeploying existing items not at the end of their useful life to less technically demanding applications. An overview of equipment available for redeployment can be kept by the clinical engineering department where it holds the organisation's equipment inventory and has a broad view of operational clinical requirements. If initial screening shows no existing equipment is available, the request then goes into the organisation's process for deciding on the priority of a purchase and justifying its funding. This process usually includes clinical staff and managers, with advice from clinical engineers,

Medical Equipment Management

and often involves a group responsible for allocating equipment budgets. The organisation's board may make the decision for high-cost or high-impact items. For large projects, a business plan from the end user will be needed to convince management and finance functions that a project is viable. The timescale from initial awareness to establishment of need may vary from a few days to a year or more.

**Example 2.1   Replacement Needs**

For a syringe driver, where technology is not changing at a rapid rate, replacement is likely to be on the grounds of physical deterioration and unreliability as the equipment comes to the end of its working life. This is often raised as a problem by nurses or doctors who suffer the consequences of unreliable operation. Replacement is also likely to involve a whole *fleet* of syringe drivers which, ideally, consists of one or two standard models. Change might also be driven by economic considerations, for example to reduce overall consumable costs.

A new ultrasound scanner might be a replacement or be intended for use in a new clinical service or function. Ultrasound is currently evolving quickly, with continuing development of new functions and improvements to image quality and portability. Equipment may be considered functionally inadequate well before it physically deteriorates far enough to require replacement on the grounds of reliability or inability to provide ongoing support.

The lifetime of an x-ray room depends on its complexity and intensity of use. A general x-ray unit can remain functional for 15 years or more with appropriate care and spare parts. More complex systems, for example, in cardiac imaging, include facilities such as electrophysiological imaging where frequent upgrades are needed to keep at the forefront of technology and effective lifetime can be 10 years or less.

## 2.3.2 Funding

Funding for low-cost medical equipment usually comes from a departmental or service budget. Medium-cost items may require allocation of funds from the organisation, and very large items often require funding to be raised externally (Chapter 5). Sources of funding include the following:

- *Revenue*: This covers items likely to be used within a financial year, such as consumables, spare parts and maintenance. UK government accounting rules for National Health Service (NHS) hospitals historically have placed equipment costing less than £5000, including taxes into this category. Other countries or organisations may use a different definition. Equipment costing over £5000 can sometimes be paid for through revenue by getting it on loan from a supplier, with its cost recouped by paying a higher price for associated consumables or by direct rental. Another way to pay for equipment via revenue is through a managed service, where all costs associated with equipment, such as consumables, maintenance and depreciation, are paid for on the basis of equipment availability or as a cost for each test or treatment.

- *Capital*: Most large organisations have an annual capital budget allocation to cover areas such as medical equipment, IT and estates, which may be augmented for specific developments in response to a good business case. Depending on the organisation's status and type, capital can be generated from income, allocated from central government monies or funded from the depreciation of existing equipment. Capital items usually have a minimum lifetime (greater than a year in the United Kingdom) and cost more than a defined amount.

- *Government or other initiatives*: Capital may be earmarked for a particular project either for a particular institution or across a whole health system. This is most often used to encourage the introduction of new technology, for example, screening for a particular medical condition.

- *Charitable donations*: These vary in size from a few tens of pounds to many millions. They are often used to provide high-profile equipment, such as new imaging technology, that would not be funded within the current health service or to support research and development work.

- *Research grants*: Another source of equipment to help develop new techniques, often treated favourably for tax purposes. Equipment purchased for research can end up providing a clinical service, as techniques gain acceptance.

- *Leasing*: Finance leases are a way of paying capital costs over an extended period, usually 5–10 years. Accounting rules vary with time and from one country to another. Leases need careful monitoring to make sure that sufficient termination notice is given and that replacement equipment is procured in time to hand equipment back, otherwise considerable extra cost may be incurred. It is worthwhile obtaining specialist advice before setting up leases.

It generally takes longer to find funding as the cost of a project increases. Processes for funding approval vary from an immediate decision by a budget manager to a prolonged and public process of consultation and decision making that may involve government intervention. Large capital commitments are usually made at the highest level of an organisation and may take a number of years to work up for complex schemes such as a new cancer centre.

Funding must also be identified to cover lifetime costs, including consumables and maintenance. This requires effective liaison between equipment users, senior management and staff in the clinical engineering, finance and procurement services. Competition to obtain funding for new and replacement medical equipment is often intense, and replacement equipment budgets are often underfunded because service need and complexity tend to grow in developing healthcare systems. Funding is less of a problem where depreciation and replacement of equipment forms part of a long-term business plan.

Individual items should be treated consistently for accounting purposes. The question as to what constitutes a single item or a capital project is open to interpretation, and an organisation needs to clarify the definitions in its standing financial instructions, whilst a purchaser must be aware of these and take care to comply with them. An item may consist of a number of interdependent units, each of which is essential for the function of the whole – for example, an image analysis workstation may comprise a computer, monitor, software and accessories, all individually costing less than the capital threshold. Where identical units are purchased at the same time, this may be considered by some organisations as evidence of their being a single capital item, particularly if they depreciate at the same rate and are scrapped at the same time. An example is a fleet of 250 syringe drivers (Figure 2.4), each costing £1000 but with a considerable combined capital value and deployed on a single site. Some organisations may capitalise this expenditure, whilst others take the view that, as items can be used and disposed of independently, they are not a single unit and hence should be revenue funded.

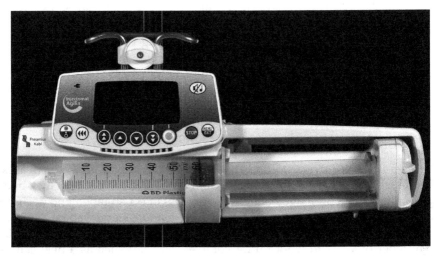

**FIGURE 2.4**
Example of a syringe driver.

**Example 2.2  Funding**

Individual syringe drivers are likely to cost less than the capital threshold, but the total cost may be large if there are hundreds of items on the inventory. This item, of the three, has the widest range of practical options for purchase arrangements which can be tied in to consumable and maintenance deals, and managed service provisions.

Ultrasound scanners are capital items, varying in cost from under £10,000 to over £100,000. Significant cost reductions are often available for multiple purchases. Continued technological innovation on these machines creates a need for regular turnover where state-of-the-art equipment is required, for example, when scanning for foetal abnormalities. This makes leasing attractive, as a machine can be leased for less than its purchase price and handed back after, say, 5 years.

An x-ray room is a capital item and is usually considered together with its supporting equipment and facilities. Funds need to be found not just for the equipment but also for any associated building works to prepare for installation, for example, to increase room cooling or add more computer cabling, and an entire room may need to be reconfigured to use equipment in the most effective way. Resources are also needed to carry out performance checks before acceptance and final payment, including radiation physicists who undertake radiation output and safety checks. New facilities require considerable expert input into their design and evaluation, to control radiation exposure and ensure supporting facilities such as power and air conditioning are adequate. A budget must be allowed for this in the project funding.

### 2.3.3  Specification

Once a proposal has funding, an appropriate model must be chosen (Chapter 6). New equipment may directly substitute for an existing make and model of device, be an improved version or introduce new technology. The more radical any change in technology, the greater the care needed to agree an effective specification and evaluate that technology in practical use. Advice from other centres which have already purchased similar technology can be particularly helpful in reducing the risk of making an inappropriate purchase, as can guidance documents from organisations and agencies that undertake impartial evaluations of medical devices (Chapter 13).

A technical specification should be developed for replacement as well as new equipment, since there may have been a change in factors such as the functionality of devices, user demands, the type of accessories and consumable design and cost. The specification defines what will make a device suitable for the intended purpose. A good specification addresses not only what is needed when the item is delivered but also tries to foresee possible changes of use during its lifetime. Standardisation is desirable for a number of reasons, including improvements in safety from user familiarity;

simplified training; economies of scale for consumables, spares and training; and a faster build-up of overall experience with the device. Equipment must support operational clinical policies and procedures and may be required to connect to and be compatible with existing devices for data transfer. The development of technical specifications for equipment used in more than one area, particularly where linked with standardisation, therefore requires an overview greater than that of any one user. This is why all purchases and specifications should be open to some level of independent review, on a scale that matches the extent of the purchase – a project group for major purchase such as a magnetic resonance imaging (MRI) scanner, a user group decision on a standard infusion device, a short discussion between an end user and clinical engineering about a blood pressure monitor or the choice of a standard model of weighing scale from a limited procurement catalogue.

There are other pitfalls in accepting non-standard equipment. A single charitable donation, however well meant, may actually drain resources from an organisation if it adds another equipment model to the existing inventory and increases maintenance, consumables and training costs. Also disproportionate costs can be incurred by local purchase of a few special discount items at the behest of an enthusiastic sales representative that then require specialised non-discounted consumables and support.

Specifications should be agreed between clinical, technical and end-user groups, to cover all their requirements. Apart from the need to consider the specification from different points of view such as function, maintenance and the costs of consumables, this also acts as a safeguard against undesirable practices that may undermine an effective, fair and open procurement process. Firstly, the specification may suffer from *function creep* as end users become aware of additional desirable functions and add-ons that would be nice to have but are not within the remit of the original project or established clinical need. It can be difficult to get end users to accept that a project allocation is agreed by the organisation on the basis of clinical need and defined functionality and is not a sum of money to be spent on buying extra features. Producing specifications also guards against the end user obtaining an initial quote at list price from the company, then later negotiating a cheaper deal and spending the money saved on something else, either to extend the scope of the initial project or to purchase something completely different. However, the organisation should not be totally inflexible and must be prepared to spend more than initially expected if, for example, additional technology does deliver business benefits, currency exchange rates fluctuate or building costs rise. The requester, likewise, must be prepared to return any savings on the initial estimate to the common equipment fund. Items that should definitely not be included in a capital purchase are maintenance costs and consumables, which should come from revenue budgets. The agreement must also not contain hidden benefits, for example, the company funding a research fellow, without these being declared and their value taken into account.

**Example 2.3   Specification**

For a syringe driver, the key specification points are suitability for use in a particular area (for example, alarm facility and size), infusion accuracy and precision. When standardising a group of devices, representative user consultation for different areas is desirable and may show more than one type of model is needed where it would be inappropriate to provide the feature on all devices in the fleet, due to cost or other issues.

For the ultrasound scanner, its imaging and measurement capabilities and transducer choice will depend on the range of intended applications, whether cardiac, obstetric, vascular, general abdominal and/or paediatric. Questions of portability might include whether equipment is to be floor standing or needs to be transported easily. A research machine might require highly specialised accessories and image processing, and a general specification will address software packages required for image storage, retrieval and processing.

For an x-ray room, a large number of items will be included in the specification, particularly for more complex applications. Accessing specialist expertise is particularly important in making sure a specification is robust, especially when purchasing is carried out by smaller organisations where x-ray rooms are bought less frequently. A number of commercial, government and professional organisations provide template specifications for different types of x-ray facility, and some also carry out evaluations of facilities in use. The team drawing up the specification will need to include experts in radiation protection, diagnostic radiology, electrical installations and facilities engineering alongside clinical users.

## 2.3.4 Tendering, Evaluation and Purchase

Many countries allow individual public healthcare organisations to choose their own medical equipment but insist they follow a tender process that gives a fair chance to prospective suppliers. Unless an item is of small value, tendering of some kind will be required by any organisation looking for value for money, and each organisation will have its own procurement requirements that also incorporate its legal responsibilities. These requirements will be set out in the organisation's standing financial instructions and will specify how many quotations or tenders are needed for a given cost of procurement. This process can range in scope from seeking a few competitive quotes to a complete tendering, evaluation and negotiation process lasting several months. The European tender process, for example, is intended to create a fair marketplace and does this through a structured process that requires inviting expressions of interest by advertisement in the *Official Journal of the European Union*, initial screening of expressions of interest for appropriateness of products and credentials of prospective suppliers, dispatch of specifications, receipt of tenders, short listing, evaluation of tenders, evaluation of the products and final selection.

Local replacement even of small items should not be organised by direct negotiation between clinical staff and a sales representative. There are many reasons for this, including breaching procurement rules and giving away an organisation's negotiating position. However good a deal seems to clinical staff, the manufacturer nearly always has the upper hand and more experience than the individual clinical user in obtaining a good outcome from a negotiation.

Framework agreements are pre-tendered and negotiated by a centralised body acting on behalf of a government or group of purchasers. This can save time and effort. With a framework agreement, the end user needs only to evaluate options available and choose a suitable model and does not need to go to tender. This does not prevent individual organisations tendering outside a framework agreement, since it is possible to achieve better value for money than a central agreement where a large replacement programme or specific needs are involved. If an organisation has itself tendered for and set up a framework agreement to define its standard model, then as long as that agreement has been reviewed and judged current, there may be no requirement to undertake additional tendering for items of the same type.

There are occasions when only one supplier is suitable. This is usually because their equipment offers specific functions not provided by another manufacturer or, most commonly, that new equipment should be compatible with existing models. Under these circumstances, it may be appropriate for the organisation to issue a waiver under its own standing financial instructions, allowing equipment to be bought from a single source. It is important to determine, however, whether the unique facilities or level of compatibility is really required, as in many applications more generic equipment is adequate. Single tender waivers are more common in areas where highly specialised equipment is used, such as research or unusual clinical procedures.

If one or more models new to the organisation are being tendered, then a documented clinical and technical evaluation by end users and technical staff is advised, to run alongside any financial negotiations as part of the tendering process. Questionnaires are a useful way to capture evaluation results and need to be drawn up in advance. A small pilot evaluation is often helpful to clarify what should be asked. Questionnaires need to be completed and evaluated objectively, with quantitative scoring included where possible. *Technical evaluation* needs to check device performance against the tender specification. Devices may well have features additional to those originally requested, and these can be taken into account in an evaluation but should be clearly identified as such. Points to be covered should include likely hazards, potential reliability, ease of decontamination, possible design weaknesses, ease and cost of servicing and extent of future service support and expected life. *Clinical evaluation* should be approached systematically, with users and areas being chosen to get a representative spread across different environments. End users will need basic training in how to use devices being evaluated in order to try them in action, so manufacturer support for this phase is vital. Key questions include suitability for each clinical application, whether there is adequate functionality, the ease of use in practice and costs of consumables. Clinical users may also spot potential hazards and design flaws.

Once all the information has been obtained, tenders are reviewed technically to ensure they include financial factors such as purchase, servicing and lifetime costs and are considered against the outcome of the technical and clinical evaluations. Organisations should check that the supplier is able to provide adequate post-sales technical and advisory support. In the United Kingdom, a pre-purchase questionnaire is available to help to acquire the necessary information, and many suppliers are used to responding to this type of query. Tender evaluation is usually carried out by a group which includes procurement, clinical users and technical specialists. There may be further discussions with potential suppliers at this stage, in order to clarify points on the tender response. A weighted scoring system is usually employed. In some countries, the outcome of formal public sector tendering is open to challenge, and a purchasing organisation needs to be able to justify its choice of final make, model and supplier. Finally, award of the tender is signed off at a level within the organisation determined by its value and standing financial instructions. Large projects—such as purchase of an x-ray set (Figure 2.5)—may require signature of the chief executive or management board and lesser ones the signature of a director such as the director of finance, down to approval by the head of supplies or head of clinical engineering. The order will be placed through a supplies or procurement function.

**FIGURE 2.5**
Example of an x-ray set.

**Example 2.4   Purchase**

Several models of syringe driver are likely to be suitable. A syringe driver has appreciable consumables costs which, over its lifetime, far outweigh the original purchase cost. The cost of a fleet of items can easily exceed tender limits, so what starts off as a small expenditure can end up requiring a full tendering process. In this case, many deals are possible – for example, a supplier may provide syringe drivers and/or their maintenance and upkeep free of charge in exchange for a contract to supply a minimum quantity of consumables per year.

For larger equipment such as an ultrasound scanner, a framework agreement may be in place and even though such items exceed tender limits, particular makes and models can be exempt from further tendering where prices were agreed through a previous competitive process.

X-ray room installation may involve tendering either separately or together for equipment and building works. *Turnkey* contracts are common, where the supplier takes responsibility for both elements and deals directly with the building contractor and equipment suppliers. During the evaluation phase of procurement, the physical size and permanence of x-ray facilities often requires visits to existing working installations by clinical and scientific members of the project team, with subsequent preparation of written evaluation reports.

## 2.3.5  Preparatory Work

Preparatory work can be a major part of a large equipping project with numerous requirements, particularly for imaging or radiotherapy equipment. Relevant questions include the following: Does the device fit into the space? Are floor loadings acceptable? Is a controlled zone around the equipment necessary, as for radiation or magnetic resonance imaging systems? How can we provide ancillary elements such as air conditioning and patient access and any special features such as radiation shielding and water treatment for dialysis? A project team will be essential to coordinate efforts by estates, contractors, end users, medical physics, clinical engineering and manufacturers, with detailed plans put in place well in advance.

**Example 2.5   Preparatory Works**

For the syringe driver replacement, suitable storage and charging facilities may already be available. However, it can never be assumed that everything is already in place simply because a similar item has been in use previously.

For an ultrasound scanner, the clinical area must be prepared, involving perhaps minor building construction and decoration, provision of waiting and reception facilities and IT support.

For the x-ray room, there will be considerable preparation, involving detailed building design and construction for access routes, checks on floor loading capacity, air conditioning and radiation shielding, together with any interlocks and safety access control. Services required will include three-phase electricity and computer network cabling to link to digital image archives. Old machinery should be removed by specialist contractors who know how to handle it, as unsecured gantry arms, toppling equipment and live electrical supplies are potentially lethal.

## 2.3.6 Delivery, Installation, Acceptance and Commissioning

### 2.3.6.1 Delivery

Except in the case of very simple devices, the delivery process requires active liaison between suppliers, end users and clinical engineering services. A paper order alone is not sufficient without discussion and verification of the required device options and accessories, the timescales involved and arrangements for installation and commissioning. Portable items are best delivered to a single point for acceptance testing, but special arrangements can be needed for large batches or for sensitive equipment. Large devices may need to be delivered to the point of installation, with very large items necessitating special arrangements including road closures, specialised lifting and handling and even building works to open up access routes. A method statement should be drawn up and consulted on to make sure all goes smoothly. Goods are too often delivered to the wrong point, resulting in lost time for those trying to trace them and sometimes permanent loss, so getting this right with the supplier saves much energy and time. A check should be made on receipt that a delivery is complete, equipment is not obviously physically damaged and all items have been received.

### 2.3.6.2 Storage

If delivered equipment is to be stored before installation or acceptance checking, this must be in an environment within acceptable ranges for temperature and humidity. For small items such as syringe drivers in cardboard boxes, a clean, dry secure area at room temperature may be all that is needed and this is true also for the ultrasound scanner, although it will require more demanding manual handling. Storage must also be secure, as criminal gangs target valuable items of medical equipment such as endoscopes that can cost tens of thousands of pounds apiece. Special conditions are required for certain consumables, such as quality control solutions for blood gas analysers, which degrade quickly unless correctly stored. Storing in corridors, other than for temporary periods during installation, is both a security and a fire evacuation risk.

### 2.3.6.3 Installation

Whilst no installation apart from unpacking is required for simple portable items, suppliers may need to carry out a long and detailed assembly and testing process on major items such as an x-ray unit. This can take days to weeks for the most complex equipment. A local site protocol needs to be agreed with the supplier, to cover liaison with local engineering staff over the supply of electricity and other basic facilities, and to ensure equipment is kept secure. Building good relationships with the installation engineer on site will facilitate the process, especially if the local scientific and technical staff is involved in acceptance testing and subsequent equipment maintenance and calibration. Less complex items such as an ultrasound scanner will have simpler set-up procedures, such as the activation of specific software options.

### 2.3.6.4 Acceptance and Commissioning

Once unpacked, the simplest medical equipment requires only a visual inspection for damage, and to check the correct items have been supplied. This applies to consumables or simple mechanical items. Generally, however, acceptance requires that supplier and purchaser run through a procedure defined by the manufacturer, to set up equipment and get it ready for testing. Additional tests may be required by the purchaser, particularly if using the equipment for a specialist purpose. Tests to establish equipment is performing correctly and safely are highly device specific. All electrical items require some functional and safety testing or inspection prior to use to avoid electrical safety risks to patients, staff and facilities. Some simple equipment may be accepted after a brief visual inspection and be switched on without any detailed tests as in the case of, say, an ultrasonic nebuliser. Out-of-hours access may be needed to set up links and interfaces to other systems, particularly when linking hardware and software to existing systems, to limit any disruption to patient information flows. Finally, equipment commissioning takes place when the end user, clinical engineer, or other party acting on behalf of the receiving organisation tests all functions in the user manual and any accessories before releasing equipment for clinical use. It is not that unusual for complex equipment to meet the manufacturer's specification but not what was agreed with the purchaser or appears in accepted guidelines, so specialist advice is essential to avoid accepting an item which is not performing to the purchase specification and to negotiate corrective action with the supplier.

---

**Example 2.6   Putting into Service**

In the case of the syringe drivers, clinical engineering or another technical service will perform any necessary functional, safety and performance tests to make sure equipment works safely. Setting up more complex equipment such as the ultrasound scanner usually requires an appropriate combination of manufacturer and in-house, clinical, scientific and technical staff, who will perform electrical safety and functional tests, enter details of the equipment onto the asset register and carry out objective image quality and accuracy checks. End users will check that clinical images are of acceptable quality and facilities and options originally specified are installed and working correctly. Reference images may be taken and kept for comparison with later images when testing for any deterioration of image quality in the long term.

At the most complex level, the x-ray room and other imaging equipment using ionising radiation will undergo long and complex operational testing by in-house or commissioned scientists, including extensive quality assurance and radiation protection measurements.

### 2.3.6.5 Payment

Payment should only be made when the organisation is satisfied that there has been a satisfactory conclusion of acceptance and commissioning. In some cases, pressure may be applied by suppliers to accept goods that are incomplete or untested, so that early payment is made. This pressure can be resisted more easily if appropriate technical expertise is available to advise the purchaser whether or not equipment is functioning correctly. Payment may be staged for a large project with a proportion being paid on delivery, a further sum on completion of installation, and the balance on final acceptance, usually after an agreed period of satisfactory operation. Staging provides the purchaser with a powerful negotiating tool should the supplier fail to provide a satisfactory working installation.

### 2.3.7 User Training

Poor training or even its complete absence is a major source of medical device risk and accounts for a substantial proportion of device-related incidents. Users fall into three main groups:

1. *Non-specialists*, who may be clinical staff using relatively simple equipment for the treatment or diagnosis of patients. These include general practitioners, nursing staff and healthcare assistants setting up devices; district nurses, physiotherapists or respiratory physiologists providing devices for use at home; and patients themselves or their carers.
2. *Specialists* such as clinical physiologists, clinical scientists and technologists using a range of complex equipment for a particular clinical diagnostic or therapeutic process, such as sonographers, cardiac electrophysiologists, anaesthetists and radiographers.
3. *Technical support staff*, such as clinical technologists and clinical scientists, who advise on and support medical equipment use.

Non-specialist users typically use a range of relatively simple equipment. They need training in how to set up and use it safely and in how to recognise and report faults. Specialised users should already be familiar with the general operating principles of equipment they use and will have background training and expertise to undertake basic troubleshooting and user maintenance. They may also use equipment under changing circumstances and need flexibility to meet new or unexpected challenges in clinical applications. Technical support staff troubleshoot faults in more detail and possibly maintain and repair the equipment. Technical and scientific operators require training where they are end users, for example, to set up, calibrate and test equipment, with additional specialist training if they are to undertake equipment repairs.

Training is a necessary part of best practice. It can be provided by the manufacturer, third-party organisations or in-house trainers. Trainers often have a clinical or technical background and may set up a cascade of training whereby lead individuals train and verify other staff members as trainers on specific devices. For governance reasons, such a training cascade should be limited where higher-risk equipment is involved and all training should be recorded and signed for by the trainee. There is increasing recognition that attendance at training is not sufficient of itself and testing individual competence is increasingly accepted as necessary, at least for high-risk equipment, both after initial training and at periodic intervals. Devices such as glucose meters are available with built-in monitoring to check an individual's competence status over a computer network and to prevent use if appropriate quality checks have not been carried out or users have not operated the equipment for some time. Chapter 7 has further details on training.

---

**Example 2.7   Training**

Users competent to operate a syringe driver will be able to demonstrate safe placing of the right type of syringes within the device, explain control functions and operate them correctly, manage battery charging and deal appropriately with malfunctions and error messages. In the case of the ultrasound scanner and x-ray facility, users are expected to be specialists in ultrasound and radiological imaging, respectively, but must still be shown how to use the controls and facilities on a particular model. Incidents still occur where individuals not trained on a specific device think they can operate it and then cause injury to patients.

---

### 2.3.8  Deployment

Where equipment is portable and has been tested in a central location such as clinical engineering, arrangements need to be made for delivery to the point of use. The timing of delivery, roll out and training and the withdrawal and collection of any existing equipment being replaced, requires extensive coordination where large numbers of items are involved. User instructions must go out with new equipment or be easily available electronically, and operating procedures such as regular daily checks and arrangements for consumables should be set up and tested in advance. A central equipment library can help with both roll-out and ongoing equipment deployment, particularly when managing large numbers of portable items such as syringe drivers, as end users return items to a central facility after use to be cleaned, charged and checked for basic operation and safety before reissue.

### 2.3.9 Asset Management and Depreciation

Records of equipment are required for two purposes: corporate governance, to cover financial and liability issues, and the practical management of devices in use for location tracking, maintenance scheduling and product recall. These two sets of records may be kept on separate databases or in a combined system with sufficient functionality. Accurate information in the asset database is needed to facilitate financial accounting and planning, including looking at long-term equipment replacement needs, tracking equipment residual value and calculating depreciation. Although the price for which a used device can be resold will vary according to its condition and the state of the second-hand market, accountants usually assume its value will fall from the purchase price to zero by equal increments over a number of years. The lifetime over which the value falls is based on estimates according to the type of equipment and may be quite different from its actual useful life: Devices can be superseded by technological advance before their intended lifetime or go on working for many years past their estimated life. The value attributed to an asset at any time is its residual value, and this can have a considerable effect on an organisation's finances. An asset may be a single device like an ultrasound scanner or a complete capital project, such as a new outpatient x-ray room and all its associated equipment.

A comprehensive equipment management database is an excellent tool for operating an in-house equipment management service. Linked records can provide a complete service history and location record which include both internal and manufacturers' service activities and schedules, along with costs of parts, sundries and labour, to help manage and maintain the equipment. Suitable data entry and information reports support day to day planned servicing and repair, as well as longer-term analysis of the cost and performance of different makes and models and the development of risk-based approaches to maintenance.

## 2.4 Management in Use

### 2.4.1 Storage

Appropriate space and conditions are needed to store both equipment and consumables. These should be planned for and justified on a risk management basis. A central equipment library can provide efficient and high-quality storage for items such as syringe drivers, supported by sufficient shelf space and battery-charging facilities in local storage areas. Portable ultrasound scanners may be no bigger than a laptop computer, so secure storage is vital, linked to location tracking to help prevent loss, theft or damage.

### 2.4.2  Decontamination: Cleaning, Disinfection and Sterilisation

The terms cleaning, disinfection and sterilisation refer to distinct processes. Cleaning is the removal of dirt and other obvious physical contaminants like blood and saline from surfaces, and its standard is judged by appearance. Disinfection is the destruction of contaminating organisms such as bacteria by chemical means. It typically takes place by wiping solutions over surfaces at room temperature and removes a large proportion of organisms. It may be part of an overall cleaning process. Sterilisation is the aggressive destruction of all organisms by processes such as exposure to steam at elevated temperature (autoclaving), immersion in noxious gases or exposure to ionising radiation. It is a complex process, requiring highly specialised equipment and scrupulous quality control systems, preceded by cleaning to remove gross physical contamination. It is used with reusable devices where infection risk is intolerable, such as surgical instruments or implantable devices, and for disposable (single-use) items.

Responsibility for routine cleaning of medical equipment during and after patient use is often poorly defined. Employees who clean floors and surfaces may not be trained in the safe handling of medical equipment, whilst clinical and technical staff may be occupied with more complex operations. The best way to avoid this problem is to establish routines for the end user which cover cleaning and disinfection, with advice from infection control staff. The latter are now very concerned with devices due to widespread hospital-acquired infections. Cleaning and disinfection should be considered as an integral part of specification and procurement, to make sure the organisation can decontaminate medical devices after purchase.

---

**Example 2.8  Disinfection**

For the syringe driver, this will involve cleaning the outside of the device with a cloth dampened with an appropriate solution, as approved by the manufacturer. Syringe drivers are particularly prone to contamination by blood and saline. For the ultrasound scanner, a similar regime is indicated, including the cleaning of stray coupling gel from the probes and the outside of the machine. Transoesophageal and other intracavity probes must be disinfected according to agreed guidelines. For the x-ray room, wipes are used to clean exposed surfaces and drapes cover equipment to prevent physical cross-contamination.

---

### 2.4.3  User Maintenance, Spares and Consumables

Users should be trained in the proper upkeep of their equipment or delegate it to a local technician or in-house department. This might include replacing plug-in leads, probes and transducers. They may also be required to carry out quality control checks including, for example, the use of special gases or standard buffer solutions for point of care testing devices or calibration phantoms for imaging systems. An adequate stock of parts and consumables for user maintenance should be provided, together with the means to keep stocks updated. Consumables need to be managed for stock levels, with regular reordering and checks on remaining shelf life. The fewer local stores that are maintained, the better, as consumables are then easier to locate and monitor. Hoarding is a common problem that leads to excessive waste, at least in the NHS [3]. Spares associated with routine maintenance and breakdown repairs need to be managed in a similar way, although the large number of possible items means that it is financially and logistically possible to maintain only the most common or essential of these as stock items. Expert advice and analysis can help when selecting spares and finding suitable alternative items.

**Example 2.9   Consumables**

For a syringe driver, giving sets are a major consumable so efficient stock-holding and storage is essential. The ultrasound scanner requires a supply of sterile sheaths if intracavity probes are used, but otherwise its main consumable is scanning gel. An interventional x-ray facility is supported by an array of items such as catheters, guide wires, stents and associated consumables. New technology can cut consumable costs, for example using bar coding to improve stock control or digital radiology to avoid x-ray film, but often requires significant capital investment and ongoing support costs.

### 2.4.4   Planned Preventative Maintenance and Breakdown Maintenance

Managing routine maintenance and breakdown repair of medical equipment is a major undertaking. Every year it typically costs between 5% and 10% of the purchase cost of equipment. Maintenance is essential to minimise the risk of failure, improve safety, maximise equipment capability and minimise unplanned downtime. It is often assumed that placing equipment on a full manufacturer's contract, renewed annually and promptly, is the way to ensure best operation and peace of mind. However, routine maintenance can be performed not only by the manufacturer but also in-house, by a third-party or by a multiservice vendor. Making the correct decision as to what routine maintenance is required for each piece of equipment, how it is to be provided, and who is to be responsible for organising and performing it can considerably improve the cost and efficiency of maintenance, improve equipment uptime and reduce clinical risk. Basing this on a risk and cost–benefit analysis is a good way forward. For example, for some items, a much faster and cheaper response can be achieved by keeping extra equipment in-house ready to swap for faulty items and to get these repaired on an ad hoc basis. Signing a service contract does not necessarily ensure trouble-free operation or outsource risk.

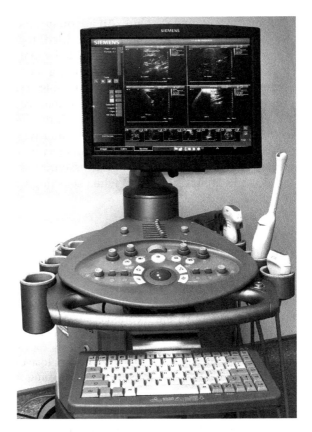

**FIGURE 2.6**
Example of an ultrasound machine.

The delivery of service contracts needs to be evaluated, managed and monitored for efficiency and effectiveness, as well as the performance of in-house or third-party services. The optimum level of maintenance and the type of service contract for the three devices we are considering depend on a number of factors, including risk and cost–benefit analysis, that will be described in more detail for maintenance arrangements in Chapter 8 and maintenance contracts in Chapter 9.

**Example 2.10   Maintenance**

All three devices require some form of maintenance. The x-ray room is most likely to be placed on a manufacturer or third-party contract, on the grounds that in-house expertise is unlikely to be available. The ultrasound scanner (Figure 2.6) may be maintained in-house if expertise is available, or through a full contract with the manufacturer, a third party or a multiservice vendor or in a cooperative contract where front-line maintenance is performed in-house and more detailed repair and service is carried out by others. These possibilities also apply to the syringe driver. For the ultrasound scanner and for the x-ray system, elements of the contract that include expensive replacement parts such as transducers or imaging panels effectively amount to a form of insurance. It is possible with an item such as the ultrasound scanner not to rely on a contract at all and to simply call out an external agency or the manufacturer in the case of a breakdown, if clinical pressures allow it.

For those items that, after a careful risk assessment, have not been placed on a service contract or given regular preventative maintenance in-house – a *fix on fail* approach – support can be given through regular inspection by end users or technical staff. Basic physical and operational checks before use do much to detect developing problems. The balance between risk, cost and benefit is instrumental in determining financial and operational efficiency. Monitoring service breakdowns and associated costs and disruption is essential to verify the adopted approach is reasonable.

### 2.4.5 Quality Control and Performance Testing

Regular performance measurement with specialised test equipment may reveal developing problems or equipment out of adjustment. This is particularly true for imaging equipment, where image quality is subject to informal clinical monitoring and frequent quality checks, and more generally for basic items where visual and inbuilt system checks can pick up developing or actual faults.

**Example 2.11   Quality Control**

Syringe driver performance checks for accuracy and linearity of delivery may be carried out annually, or after repairs or incidents. Ultrasound reference images may be taken using test objects or well-defined anatomical views of which the end user has expert knowledge. Quality assurance checks may be carried out annually by the operators by clinical engineering, or there may simply be reliance on an annual manufacturer's service visit. For the x-ray room, with its large patient workload, any breakdown and consequent downtime have a considerable effect on clinical service provision and regular quality monitoring and testing after maintenance is therefore of fundamental importance. Compliance with ionising radiation regulations and clinical governance also demands that radiation exposure and image quality are regularly checked to make sure they fall within acceptable limits. This is important for digital systems, where image manipulation by operators post-exposure, to get the best results, can mask incipient problems.

Entry of equipment details onto a medical equipment database is of particular value. In addition to routine quality testing, records covering frequency of breakdown, maintenance and training resource costs, consumables consumption and effective life then support analysis and calculation of costs, downtime and other relevant performance measures. Clinical engineering is then in a position to advise when best to replace equipment, identify any need for further user training and suggest changes to maintenance and management arrangements, through use of an available evidence base.

Equipment models can gain a reputation for unreliability or ineffectiveness, whether justified or not. Some equipment regularly turns up for repair with no fault found. Sometimes clinical incidents are due to poor training but are blamed on poor equipment. Experience shows that users who enthusiastically chose a model at the evaluation stage may be the first to condemn it and demand its immediate replacement when problems arise. The engineer may encounter a situation where the technical, or training, solution to a problem is theoretically easy but the culture of hostility towards the equipment is so developed that there is no choice but to replace it. This represents a failure of the equipment management system.

---

**Example 2.12   Breakdown**

Syringe drivers can be particularly prone to failures and incidents due to poor user training, inadequate battery charging and mechanical mishandling. A medical equipment library, coupled with regular top-up training sessions, is an excellent proactive remedy.

The ultrasound scanner can be monitored for deteriorating performance by routine imaging checks and by monitoring the number of breakdown call-outs and lifetime of transducers. Such information may show that changes in the maintenance regime and the means adopted for probe replacement, which could be a contingency fund or comprehensive maintenance contract, may need review. Excessive damage to probes due to mishandling can be identified and remedied by training. Older scanners are sometimes retained for *emergency* use in out of the way locations, but this is almost certain to be counterproductive unless maintenance and training are kept up. It is asking a lot of on-call staff in an emergency to get good results from an old and unfamiliar machine with a performance likely to have deteriorated due to lack of use or maintenance.

The reliability of x-ray equipment is affected by its pattern of use. For example, x-ray CT tubes fail earlier if they are used infrequently and without the proper warm-up cycle. An x-ray system is such a large investment that even when a system is shown to be performing poorly, there is a significant barrier to shutting it down and replacing it. Organisations need to be alert for this and be prepared to find funding for replacement systems in advance, so that patients do not receive higher radiation doses or clinicians try to diagnose from poorer images.

### 2.4.6 Condemning and Disposal

Equipment judged to be beyond its useful working life in a particular application should be reviewed for condemning or redeployment. Good financial governance requires that the decision to condemn or dispose of equipment is made by properly qualified staff, in a transparent way, with appropriate records kept. This guards against, for example, fraudulent condemnation and resale of good equipment. The end user, together with clinical engineering and finance departments, should be involved in an agreed process that is performed according to set procedures. It is also important to check actual ownership of items before disposing of them, as there have been cases in the authors' knowledge where organisations have inadvertently disposed of leased and loaned items and have then been required to compensate the legal owner.

Disposal can be by scrapping, by sale or by donation for reuse after refurbishment. Particular considerations mean that leased equipment requires more effort to coordinate its disposal and replacement. An organisation has to make sure that items are returned on time at the end of the leasing period to avoid paying excessive rental charges, with an end of life inspection to fit in and need to put replacement equipment into service ahead of the return date. Since in practice most equipment leases are extended or otherwise overrun, it is vital to negotiate how such an overrun will be handled when a lease is set up, to minimise future costs. Leased equipment also needs to be returned with all its accessories, including manuals. Charges can be and are levied for damaged or missing items.

Where laws exist to require disposal and recycling by the manufacturer, as in the European WEEE Directive [4], equipment can be returned to the manufacturer for scrapping, provided that they are still trading. However, charges may legitimately be levied by the manufacturer unless there is a contract to the contrary in place. Older equipment may be scrapped and disposed of by the organisation, via specialist contractors if necessary. This can be particularly costly for items containing radioactive sources or other hazardous material. Companies exist that specialise in the retrieval and refurbishment of equipment for sale, often by auction. Both the syringe driver and ultrasound scanner might be disposed of by this route, and it is possible to have an arrangement whereby all surplus and broken medical equipment for a healthcare organisation is disposed of via a single company. Specialist suppliers also exist who purchase and refurbish x-ray facilities. Chapter 12 has more information on this topic.

## 2.5 What Is Clinical Engineering?

*Clinical engineering* is a professional activity centred on the design, development, support and management of medical devices. Related areas of interest include biomechanics, biomaterials and mathematical modelling of physiological systems, and these are considered by some groups to fall under a wider definition of *medical engineering*. In the broadest sense, a clinical engineer is concerned with application of the physical sciences to medicine, through engineering. This is in contrast to a biochemist or biologist who applies the biological sciences to medicine. Where this activity takes place in a university, the discipline is usually known as biomedical engineering. However, where the engineer is working in a hospital or community context applying the discipline to patient care, it is known as clinical engineering.

Medical equipment management is an essential subspeciality of clinical engineering. The group within a healthcare organisation that deals with the technical and scientific aspects of medical equipment management may variously be named clinical engineering, electromedical engineering (EME), electrical biomedical engineering (EBME) or biomedical engineering (BME). Clinical engineering services are often managed in hospital as part of a group of clinical support services together with imaging and pathology, or can be included within an estate and facilities function, form part of an academic institution or be provided by a commercial or not for profit company. An in-depth review of clinical engineering worldwide is given in [5]. Appendix A describes practical aspects of running an in-house equipment management service with particular reference to the clinical engineering department.

## 2.6 Summary

In this chapter, we have followed three items of equipment through the equipment management process within an organisation. These three devices take very different inputs by way of skills, physical and financial resources and time to manage through the process, both quantitatively and qualitatively, from demonstration of need, procurement, commissioning, deployment, maintenance, monitoring and ultimately disposal. Carried out well, these processes will minimise risks from operating medical equipment and are described in further detail in later chapters.

## References

1. European Commission. The Medical Devices Directive 2007/47/EC, pp. 21–55. *Official J. Eur. Union*, L 247/21 (accessed on September 21, 2007).
2. European Commission. *MEDDEV 2.1/6* – Guidelines on the qualification and classification of stand-alone software used in healthcare within the regulatory framework of medical devices, 2012. http://ec.europa.eu/health/medical-devices/files/meddev/2_1_6_ol_en.pdf (accessed on August 11, 2013).

3. Goods for Your Health. *Improving Supplies Management in NHS Trusts.* Audit Commission, London, U.K., 1996.
4. European Commission. Waste Electrical and Electronic Equipment (WEEE) Directive 2012/19/EU, pp. 38–71. *Official J. Eur. Union,* L 197/55 (accessed on July 24, 2012).
5. Dyro, J. *The Clinical Engineering Handbook.* Academic Press, London, U.K., 2004.

# 3

## Medical Device Risk, Regulation and Governance: An Overview

### 3.1 Introduction

Healthcare organisations face multiple regulations and legal liabilities that affect how they manage their medical equipment. Government and regulatory bodies expect compliance with relevant legislation, regulations and standards, whilst professional bodies encourage best practice. At their best, these requirements are intended to reduce the overall level of risk to patients, staff, the public and the organisation. Reducing the number of incidents and improving system performance benefit patients, improve efficiency and release resources for patient care. In the worst case, serious incidents can have significant financial consequences, and the size of any financial or other penalty is often directly related to how well risk has been managed.

In this chapter, we introduce risk management concepts and consider their application to equipment management and clinical engineering, outlining examples and highlighting issues raised in later specialised chapters and appendices. First, we give an idea of the frequency, consequences and costs of medical device risks. We then take the equipment management life cycle described in the previous chapter, highlight some of the major risks arising at each stage, and identify practical steps to minimise their effect. We describe in some detail the principles and processes of risk assessment and risk management, including identifying risks and devising and monitoring schemes to reduce or eliminate their effects. We consider physical risks from electrical, radiation and infection hazards in addition to systemic factors posed by inadequate training and weaknesses in the way processes are managed and monitored.

Thoughtful compliance with corporate governance requirements and standards of good practice is likely to improve patient safety. Following good best practice may also reduce any damages awarded in the event that a harmful incident occurs. We explore these concepts and their contribution to reducing the cost of purchasing indemnity against civil damages claims.

In the last part of this chapter, we describe some major legal obligations facing an organisation which provides clinical services, particularly those regarding the safety of medical devices themselves and health and safety issues surrounding their use and technical support. We differentiate between legislation, regulations, standards and best practice and introduce examples of each. To do this, we provide an introduction to specific legislation relating to the manufacture and management of medical devices, and also to the general civil and criminal law concerning consumer protection where it relates to medical devices. Alongside this legal framework, we introduce some of the regulations and standards relating to medical devices. Health and safety legislation is important in the physical manufacture, handling, care and maintenance of medical devices and its key points will be introduced here. Finally, we refer to the types of health and safety law which have the greatest impact on medical device management. Although legislative approaches to risk vary between national legal systems, the risks themselves remain the same, and by highlighting United Kingdom and European law, we aim to give the reader a base from which to review the effectiveness of their own national legal framework.

## 3.2 Medical Device Risks

### 3.2.1 Frequency

The actual level of risk from medical devices can only be estimated because there is good evidence that only a minority of incidents are ever reported. A comprehensive US study [1] in 2000 estimated there were 454,000 medical device-related incidents nationwide, four times the number actually reported to US regulatory bodies. A comparable proportion of unreported incidents is seen in the United Kingdom, where in 2008, for example, out of 920,000 incidents reported in the National Health Service (NHS) in England and Wales, 27,720 or 3% were related to medical devices [2] yet only 8,900 were notified directly to the body which monitors device incidents in the United Kingdom, the Medicine and Healthcare products Regulatory Agency (MHRA).

Perhaps a better measure of the scale of equipment-related risk can be obtained by looking at the reporting of more serious events. A quarter of the 27,720 medical device incidents reported in 2008, a total of 6,700, allegedly involved some degree of patient harm with 220 resulting in severe injury or death. MHRA data that year show they had received reports of 212 deaths and 1200 serious injuries and investigated 190 fatal incidents, but only in 44 of these could a direct link be established between the device and the cause of death. Most MHRA investigations were into four high-risk areas – implants, surgery, dialysis and life support equipment – supporting the idea that serious incidents are well reported but there is significant under-reporting of less serious events. These statistics also illustrate how difficult it is to compare data from reports made through different reporting systems using different category definitions. Extracting figures for more specific types of incident is equally difficult, as even common classification criteria can be interpreted differently by individuals who report incidents. Common categories include unavailability of equipment and user errors, but the most frequently recorded category is usually unknown. This points to causes beyond breakdown or equipment damage and suggests that failure to manage user training properly can be a major cause of risk to patients.

When looked at in the context of an overall health system, the total number of serious injuries or deaths directly attributable to medical devices is relatively low. For the NHS in 2008, there was a total of 12,000 deaths and severe injuries from causes other than the 44 attributed to medical devices, in a health system where approximately one million interventions are carried out each day. The three most common events were problems with treatment procedures, patient accidents and infection control. Equipment related incidents can and do occur in any healthcare environment.

It is much more difficult to establish the number of times medical device problems contribute to a reduced quality of patient care or increased costs. Shortcomings such as a lack of suitable equipment, poor device design and inadequate user training can frequently be inferred from the content of incident reports submitted by clinical staff and from anecdotal evidence. The direct and indirect costs of equipment management problems are likely, therefore, to be highly significant.

### 3.2.2 Legal and Financial Consequences

Should a patient come to harm, either by being injured or failing to receive appropriate treatment through medical equipment failure or non-availability, the healthcare organisation may be liable to a civil claim for compensation. Similarly, if staff working on equipment are injured, the organisation may face both civil and criminal liabilities from failure to comply with health and safety law. Failure to comply with some aspects of medical device legislation is in itself an offence liable to criminal penalties.

The total current and future cost of litigation to the UK NHS is estimated to be in the region of several billion pounds (and rising) out of a total budget of about £100 billion. The costs of individual actions are significant: in 2012-13, 10,129 claims of clinical negligence against NHS bodies were received by the NHS Litigation Authority, the NHS pooled insurance scheme, and £1.1 billion was paid out in damages and costs with an estimated total of £23 billion future liabilities [3]. These amounts will be only a proportion of the overall UK total, due to the involvement of other insurers or direct payments from healthcare organisations.

The contribution medical devices incidents make to overall NHS litigation costs is not readily available, particularly as information on out-of-court settlements is unavailable. However, given the number of serious incidents, and extrapolating the proportion of medical device to clinical incidents, suggests medical device problems cost the NHS over £20 million a year.

## 3.3 Risk Management

In this section, we look at the ways in which risks can be assessed and managed.

### 3.3.1 Risk Categories

The three major categories of risk challenging a healthcare organisation are clinical, health and safety and corporate.

*Clinical* risks typically relate to sub-optimal patient outcomes such as misdiagnosis, inappropriate treatment, acquired infections or adverse drug reactions. Whilst most medical equipment incidents are minor, they can cause serious patient injury or death. A detailed analysis of equipment-related injuries in one hospital showed that the most common recorded cause was human error [4].

*Health and safety* risks expose patients, staff and the public to potential injury from equipment, materials or the environment. Measures to control these risks often result in national legislation and local inspection.

*Corporate (organisational)* risks are often linked to those from clinical or health and safety causes and might lead to loss of business, waste of financial or other resources or damage to the organisation's reputation. Such risks, if serious, may even threaten its continued viability, as in the case of natural or civil disasters.

In the clinical environment, patient, client and staff safety is a priority. Threats to safety include the improper operation, incorrect setting up or unavailability of functioning diagnostic and therapeutic equipment, in addition to any harm associated with equipment malfunction or poor maintenance. Safety failures create unnecessary suffering and expose organisations to civil action for compensatory damages and also to criminal action which may result in fines or imprisonment. Actions may be directed against an organisation and at individuals within it such as the chief executive and any clinical staff directly concerned [5]. Adverse publicity may then lead to loss of business and income. Another cost is the time spent by the organisation on investigating and processing incidents, but if done effectively any consequent policy review and corrective action is likely to improve patient care and safety in the long term.

The impact of regulation on medical equipment management is growing, in line with society's greater perceived readiness to engage in litigation. Arrangements that were once widespread and seen as beneficial, such as loaning medical devices between healthcare organisations and the use of surplus equipment for research, are now fraught with both actual and perceived legal complexities and risks. In such a climate, it pays to manage risks and not ignore them.

## 3.3.2 Perception of Risk

The aim of medical equipment risk management is to keep patients safe whilst enabling staff to carry out effective diagnostic or therapeutic procedures. This means equipment must be safe and effective in the clinical environment with procedures carried out by staff trained to use that equipment, whether for a routine intervention or a clinical innovation. If only it were that simple!

Many procedures using medical equipment are inherently risky or unavoidably put patients at risk from errors, accidents or unexpected failures. Reducing these risks is usually a compromise between what is desirable and what is achievable within finite time and resources. Patients, clinical professionals, managers, politicians and society each play a part in setting the relative value placed on different outcomes and the cost of various actions to alleviate risk. A clinical engineer has to work with these perceptions and concerns, as expressed directly and also indirectly through legislation and guidance. The value of a clinical engineering professional in this context is their ability to provide a sense of proportion, based on expert knowledge and experience, when evaluating the nature of risks and benefits from a technical perspective. It is easy to be risk averse, as anyone should be when acting outside their area of expertise; however, the equipment management professional earns their salary by advising their clinical colleagues and their organisation how to minimise risk in a cost and resource effective way with minimal restrictions on essential activities. Part of carrying out this advisory role effectively is an ability to understand the concerns of other groups involved in using or controlling medical devices so that ideas and conclusions can be expressed clearly and understood appropriately.

A number of factors can distort the management of risk and need to be resisted if resources are to be used efficiently. In a healthcare context, stories of individual clinical incidents can be very powerful in raising the profile of a particular concern and distorting perceived priorities. These include political and public perceptions of risk and the culture within the organisation.

Some organisations are so risk averse that they rigidly pursue an absolute compliance with regulations, often applied in areas that they were not originally intended to govern. It is difficult to balance risks in an environment where different regulatory bodies insist on absolute compliance, without diverting resources away from real risks to hypothetical threats.

The perception of statistical risk is also often distorted, with the likelihood of extreme hazards overestimated and risks from more common events underestimated [6]. Often an issue perceived by the press and public to be highly risky is an item featured in a recent scare story, leading organisations to fear the effects of adverse publicity. Safety, in its naïve interpretation, is the absence of risk, and whilst this is impossible, the perception of safety may be principally a matter of emotional reassurance.

### 3.3.3 Practical Approaches to Risk Management

The management of risk involves taking actions to reduce the likelihood or severity of a negative event to an acceptable level, within a level of resource proportional to the level of risk. It is sometimes possible to remove a risk entirely by stopping an activity or changing the procedures or devices used. Equipment-related risks might be reduced by measures such as special training, placing a physical distance between the process and workers, restricting access or the use of protective clothing or equipment. However, no process can be made entirely risk-free.

One system for estimating and quantitatively assessing risk uses a combination of the likelihood of an event occurring, and its consequence, to assign a combined risk score [7]. Chapter 5 contains a summary of this approach, which aids risk management but cannot take account of every issue. It involves a degree of subjective judgement but is good at ranking risks and guiding where effort should be directed to reduce them. It also provides evidence of decision making that can be audited and scrutinised externally. This technique looks only at identified threats, so a thorough risk assessment must be performed initially to consider all possible risks before scoring them and prioritising practical risk reduction measures. Prior experience provides a reality check: Where an organisation has been carrying out an activity without incident for many years, then the likelihood of any hypothetical risk is likely to be low, unless it is completely new.

Figure 3.1 summarises the different stages of the risk management process, which are as follows:

- Set out the *scope* of what a risk assessment will cover – the areas or procedures involved.
- Establish *awareness* of risks, by analysing an incident or statistical data or by thinking about what might go wrong.
- *Record* the risks onto a register and communicate serious risks to the organisation.
- *Evaluate* risks, their nature, possible ill effects and the likelihood of each one happening.
- *Prioritise* risks: the order they should be tackled in and those which are urgent compared to ones already on the risk register.

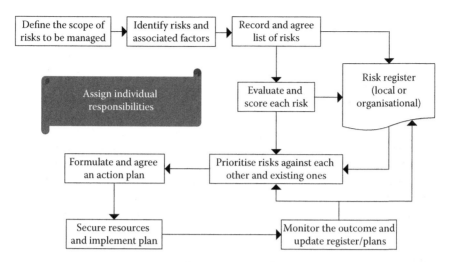

**FIGURE 3.1**
Stages of the risk management process.

- *Formulate an action plan* to address the risks and identify what resources are required.
- *Secure resources* to put the action plan into effect.
- *Monitor* whether actions have reduced risk by looking at outcomes and staff attitudes.

An organisation should have clearly defined routes for reporting risks and incidents, and to assign responsibility to individuals for managing risks and monitoring the results. A central risk register is a good way to communicate risks and their severity to the organisation, and to raise awareness of more serious risks and the resources needed to tackle them.

Risk management can be surprisingly counter-intuitive. Procedures that seem very hazardous can, with the right control measures, be made relatively safe. For example, consistent and long-term efforts to reduce risks in nuclear power generation and aviation have made them safer for those involved than other forms of power generation or transport. These lessons are being applied to healthcare, which is why reporting incidents and near misses is so important.

Safety is enhanced by meeting standards and reducing risks. The general principle applied in the UK law is to reduce risks to be as low as reasonably practicable – the ALARP principle. The term *practicable* implies that the effort made to reduce a particular risk should be appropriate within a given context, when political and financial factors, resource limitations and possible adverse consequences caused by risk abatement measures are taken into account. An organisation can only judge the appropriate level of resource to put into managing a particular risk by first making an assessment of its severity and then putting it into the context of all the risks that organisation faces. Hence, a series of risk assessments need to be made and brought together into an overall risk register, which compares both initial risks and how they might be reduced by a series of different control measures. An informed decision on where to concentrate effort and resources can then be made across the whole array of risks faced by the organisation. This whole systems approach to risk management requires a clear overview, the ability to obtain and integrate information from a range of specialist advisers, and a decision-making structure that is informed and aware of the risks concerned. Most large organisations will go some way towards achieving an integrated approach but face significant problems keeping abreast of changing priorities. In practice, every area – including clinical engineering – will need to develop its own risk register, and the organisation will need an effective forum in which very disparate risks can be weighed and prioritised against each other. Some classes of risk arise only at organisational level, such as the impact of a failure of regulation, and clinical engineering will be able to advise on any such risks in medical equipment use and management.

### 3.3.4 Risks in the Hospital Context

In contrast with generally held expectations, a hospital is a relatively unsafe place to be. One in ten patients is estimated to suffer some kind of adverse event during a spell of inpatient care [8], with injuries most commonly due to equipment misuse or lack of understanding rather than equipment failure. For example, infusion devices are associated with a substantial proportion of device-related incidents (12% of those reported to the UK regulator in 2009), yet most injuries appear to be due to inadequate user training or user error, through mistakes such as incorrect dose calculations or errors in setting pump rate. Even when hazards and the ways of avoiding them are well known, individuals may still make errors which lead to injury, such as diathermy burns during surgery due to poor electrode placement. Familiarity is a major factor leading to the under-appreciation of risk: for example, wheelchairs might be considered a safe medical device and their use and handling appears obvious, yet they are a significant cause of injury and death accounting for over 10% of medical device incidents reported to the UK MHRA.

Given the potential level of clinical risk in a healthcare setting, and the many different contributing factors, how should an organisation use its clinical engineering department to reduce medical device risks? The broad framework of such a topic-based risk management process includes the following steps:

1. Consult relevant regulatory and professional guidance and standards.
2. Identify and assess major medical device risks in the local environment.

3. Identify steps already being taken to ameliorate these risks and investigate which actions would reduce them further.

4. Develop and seek organisational approval for action plans to reduce risk.

The various factors discussed earlier then become important in deciding which actions have the highest priority. It may be that the level of resources put into treating some risks is disproportionately high and therefore that reduced effort can be justified. Routine electrical safety testing is a good example. Medical equipment differs from domestic and industrial equipment, as its technical design has to cope with potential hazards in the clinical environment including injury to vulnerable patients from low-level leakage currents. Connecting a computer to a medical system, to record data for later analysis, can create significant dangers to patients from microshocks and earth leakage currents (see Appendix B). Testing new configurations of equipment is therefore essential, whereas regular testing of medical equipment, whilst almost universally practiced as meeting statutory requirements, has failed to demonstrate widespread deterioration of equipment safety in practice. Consideration of the way failures occur suggests a mixture of visual inspection and instrumented testing provides a more cost-effective approach to risk reduction than blanket testing. For example, a survey by a London teaching hospital Clinical Engineering Department in 1997 analysed 20 years of retrospective results from 25,000 medical-grade safety tests on portable medical equipment and 12,000 on portable non-medical hospital equipment found in the hospital [9]. These equipment items included a substantial legacy of equipment introduced before comprehensive equipment management procedures such as universal acceptance testing and CE marking was introduced and hence represents the highest level of non-compliance likely to be found. The results were revealing:

- 5% of equipment was found to present a *potential* hazard, due mostly to unreported mechanical damage or because it had not been tested initially having circumvented normal acceptance procedures.
- Ten percent of this potentially hazardous equipment (0.5% of the overall total) presented an *actual* hazard that could have caused injury.
- Ninety-five percent of these actual hazards were detected by simple visual inspection, the remaining 5% (0.025% of the overall total, or nine items in all) being identified by a simple earth bond test.

This internal report concluded that the two principal methods to assure ongoing electrical safety were prompt reporting of damage by users and strict adherence to acceptance procedures. The organisation subsequently replaced routine electrical safety testing by visual inspection for most categories of equipment, retaining instrumented testing only for certain categories of equipment where a risk assessment showed this to be important.

A second area to concentrate on is front-line user support. A reasonable rule of thumb used by clinical engineering departments is that a third or more of *equipment not working* requests for technical support are caused by user problems. Where users are unfamiliar with equipment, there are obvious additional risks to patient safety, so it can be argued that concentrating on user education and training will address not only the first third of unknown problems but will, in the long term, significantly improve patient safety and care. Responsibility for the safe operation of devices, for pre-use checks and for getting trained lies squarely with the user, and no amount of planned maintenance or testing can prevent damage or danger due to misuse. That is why clinical engineering departments should offer in-depth user support from engineers or technicians with good analytical and communications skills.

Finally, clinical engineers should actively seek to reduce risks. The most effective way is to work closely in a team with end users, collecting evidence by observing and analysing procedures and evolving better ways of working across the whole system. Engineers need to beware of jumping to technical solutions, by incrementally improving designs or coming up with new devices or additional functions, but should take a whole systems overview and bring a broad range of skills to bear including data analysis, ergonomics and human factors research [10]. Developing these skills is certainly more challenging than carrying out routine safety checks, particularly when this requires negotiating with and training people who are unlikely to see their role as being concerned with equipment management and safety, but the consequent improvements can be much greater.

## 3.4 Governance, Standards and Best Practice

### 3.4.1 Introduction

Good healthcare organisations strive continuously to improve the quality of care they provide. This includes delivering better treatments and reducing risks and threats to patient safety. They work consistently to improve processes and procedures, using data monitoring and analysis to guide the introduction of incremental change. Excellent organisations learn from the example of others and continuously seek to improve the services they offer, building on and stretching existing standards. What is perhaps the most critical step towards excellence is an emphasis on tackling problems using a whole systems approach – not just looking at specific actions to improve quality and reduce risk but doing so with all those involved in every aspect of the service. In this overview, we introduce the principal concepts and systems for delivering clinical governance and examine some practical problems. More information regarding clinical governance is available in textbooks on the subject [11].

### 3.4.2 Clinical Governance

*Governance* is a term referring to the laws, policies and structures that affect the way people control, administer and are held accountable for various aspects of an organisation. Governance systems and processes lead, direct and control organisational functions. Financial governance is perhaps the most familiar area, with the expectation that, in return for being granted autonomy and other privileges by its owners and the state, an organisation will be honest and accountable for the way it manages, reports on and accounts for financial transactions and other resources.

Healthcare organisations have to deliver not only financial and organisational objectives but also clinical services to meet patient safety standards and markers of clinical quality. The delivery of clinical services is thus subject to governance processes as well. Clinical governance is the term describing those systems through which healthcare organisations seek to control and are held accountable for the clinical quality of their activities – including the management of clinical risks. Underlying its introduction is the observation, based on work in different countries and sectors, that taking a systematic approach to reducing risk and improving the safety and quality of care is much more effective than making piecemeal improvements, resulting in a greater reduction in the occurrence of avoidable incidents [12].

The concept of clinical governance started developing centrally in the United Kingdom in 1997 following an intervention by the UK's Chief Medical Officer [2]: *'In 1997 I drew attention to the fact that quality of health care did not seem to be as high on the agenda of the NHS as, say, financial and workload targets. Yet quality was what mattered to most patients, doctors, nurses and other health professional staff'*. In 1998, the Department of Health formally defined *clinical governance* as [13]: *'a framework through which NHS organisations are accountable for continuously improving the quality of their services and safeguarding high standards of care, by creating an environment in which clinical excellence will flourish'*. The risk, regulation and governance framework of the UK NHS then developed rapidly, and by 2002, clinical governance structures were largely in place with individual organisations embedding these within their own processes and systems.

Clinical governance covers the various principles and processes that allow a healthcare organisation to measure and improve the quality of its clinical services. It relies on a range of measures such as the following:

- *Introducing a culture of learning* to support continuous improvements in practice and safety. This will involve both formal individual training and organisational learning from areas such as incident investigations, risk analysis and clinical risk management, leading to continuous improvement based on research and observation.
- *Rapid introduction of good clinical practice and safety measures* found to be effective elsewhere, for example, the implementation of evidence-based guidance and standards.

- *Clear lines of accountability for quality* within an organisation, running all the way from individual staff up to each organisation's Governing Board.
- *Establishing and maintaining effective communication routes* for governance issues.
- *The provision of quality assurance monitoring and analysis tools,* ranging from individual annual appraisal for clinical staff to systematic clinical audit of actual performance against expected standards and benchmarks.

Every healthcare organisation should have a policy and management framework setting out responsibilities for implementing safety, audit and reporting measures across all its clinical activities. The outcome of clinical audit and governance monitoring from all wards and clinical departments is collected through a clear management system and reported to a clinical governance oversight committee under the Governing Board. The Board in turn oversees external reporting, which will provide information not only to government on progress against priorities and key targets but also to independent regulators and observers of clinical performance. Medical device issues are likely to be raised through this mechanism, with clinical engineers attending the oversight committee periodically to discuss incidents and present an annual report on device risks.

For an organisation to improve its relative performance, it needs to look outward as well as inward, as discussed in Chapter 14. Clinical performance needs testing and benchmarking against outside comparators and by external assessment. Various organisations design, develop and promote clinical quality performance indicators, but for these to be effective in improving local services, they must address real-world problems which are relevant to an organisation's clinical practice. There is an increasing and welcome trend to involve patients in developing suitable measures, not only through surveys and other forms of feedback but also in direct participation with clinical staff in service improvement initiatives.

### 3.4.3  Quality Systems, Records and Document Control

Quality as a general concept describes how well an organisation's services or products meet user needs. For an organisation, quality means getting it *right first time,* correctly identifying what a user or client wants and delivering it with the minimum cost in time and resources. Many techniques have been developed to help achieve this, ranging from customer surveys and process monitoring to integrated approaches such as *Six Sigma* and *Lean* improvement techniques [14]. From the point of view of the customer, *quality* means two things: a product or service that does what they want it to do, and the reassurance that this product or service will be reliable and effective for as long as they need it. The first can be seen or experienced but the second – reassurance – is far more difficult to create and sustain.

Organisations can provide their customers with assurance about the quality of their services by taking two steps. The first is to define a set of quality standards which demonstrate their commitment to meeting customer requirements. Typical elements include setting up regular review of customer expectations and needs, monitoring performance against self-defined or external standards, providing appropriate traceability for components and consumables, dealing with nonconformities in services or products, auditing performance and compliance with quality standards, and taking preventive and corrective action. The second step is to integrate all these elements into a quality system, which contains policies, procedures, forms and other documentation which staff work with to deliver a service to the agreed standard. National and international standards such as International Organization for Standardization (ISO) 9001 and ISO 13485 (see Chapter 13) set quality system elements in a proven framework aimed at continuous improvement, and Figure 3.2 provides a generic overview of how these multiple factors are brought together. Experience suggests that the following features help a quality management system add value to the work of an organisation:

- Senior management committed to improving quality and using quality management structures and techniques

- A quality system which is flexible and open to change, where it is easy to make multiple minor improvements and innovations that improve quality

- Building inspection into every stage of a process rather than inspecting only its outcomes

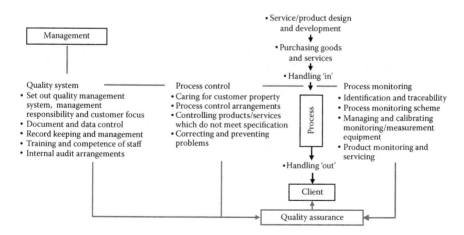

**FIGURE 3.2**
Generic elements of a quality management system.

- A focus on quality of outcome rather than following detailed and rigid procedures
- Auditors who suggest improvements as well as look for failures
- Audit reports reviewed by management and improvements made immediately
- External certification and audit to keep the organisation focused on improvement
- Staff encouraged to analyse current performance and seek improvements
- Working to improve quality in collaboration with suppliers, staff and customers

There is an extensive literature on introducing and running quality management systems. Their value is limited if they are seen as a way to do the same thing more consistently and is greatest where they are integrated into an organisation as a tool to help it learn and improve. Registration with an external quality monitoring organisation for compliance to an international quality standard such as ISO 9000 is further reassurance to customers that the quality system itself is robust and effective. Formal certification is increasingly a requirement for clinical engineering services and is almost essential to meet some legislative requirements where equipment is developed or modified in-house. In Europe, the ability to CE mark products manufactured or modified in-house depends critically upon achieving appropriate quality registration for the relevant manufacturing functions (see Chapter 13). Certification also demonstrates that an organisation is meeting many of the aspects of good governance, including audit and improvement. It also provides value in a competitive market. Embedding performance measures such as customer surveys or relative cost effectiveness in a quality system (see Chapter 14) can demonstrate the value of a service to its users. Measures vary from the seemingly quantitative and objective, which might include relative cost or value for money, through to the qualitative and subjective such as fitness for purpose and the effectiveness of safety and risk management. They should be chosen carefully to reduce the effort required to collect them and to make sure they provide a good basis for triggering corrective action or changing underlying processes.

Record keeping is central to clinical and financial governance. Well-managed records support the organisation in monitoring and analysing its performance and demonstrating it internally and externally. Records also provide an audit trail that in equipment management can show how an item has been maintained and that essential upgrades have been carried out. A fine line has to be trodden between a defensive approach which demands comprehensive records – 'if you didn't write it down you didn't do it' – and a practical approach to limit the amount of data gathered. Conformance with a quality system such as ISO 9000 series standards does not necessarily demand extensive records, as long as there is a well-designed procedure in place together with an integrated means of tracking that it has been followed. A signed tick box on a paper record, or a dated entry in a computer application or an e-mail, can all provide sufficient evidence for audit purposes.

Document control is another important factor addressed by a quality system. Ensuring the latest versions of procedures and protocols are in use requires discipline and tight version control. Serious clinical incidents have arisen where staff have printed out and continued to use old versions of updated documents for convenience, and in areas such as radiotherapy there has to be continued vigilance to avoid this happening. Electronic document control systems are available and can be used either stand-alone or as a component of a wider quality system which also sets out procedures for communicating with staff about updates and for removing superseded documentation.

## 3.5 Risk Management and Governance in the Equipment Life Cycle

Risk management and governance concerns arise at every stage of the equipment management life cycle, and this section looks at the most important areas. Legal liabilities facing the organisation must be borne in mind at each stage, including the need to indemnify against loss.

### 3.5.1 Procurement

Procurement includes all routes by which medical equipment becomes the responsibility of a healthcare organisation. The most straightforward to manage from a governance point of view is outright purchase, including purchase from donated funds if these come with no conditions attached. Equipment can also fall under the governance of an organisation through prior arrangement (leasing, hire, trials, loan, rental, direct donation or research) or when it starts using it by default. Actual examples of the latter include a patient arriving already connected to devices belonging to another organisation, a machine originally purchased for research which starts being used in a clinical service, and a surgeon who purchased their own equipment and used it on the organisation's patients. It is important under these circumstances to clarify who is responsible for the different elements of equipment management at the earliest opportunity. Indemnity and allocation of responsibility for devices on loan is an important part of medical device management, particularly for training, repair and servicing. Where equipment is on trial or loan, an indemnity from the manufacturer or supplier is required to cover claims that fall into four categories – defective design, defective manufacture, inadequate warnings and negligent surveillance or notification of potential problems. Industry or national agreements or forms may be available for specific devices.

Any organisation has to prioritise its spending on medical equipment. Patients are exposed to significant clinical risk if obsolete, inappropriate, or unreliable equipment is used in their treatment. If existing equipment cannot provide minimum accepted standards of diagnosis or treatment, a need for replacement is clear. Greater judgement is required when considering whether performance is good enough when clinical standards keep rising, and yesterday's technique has been replaced by what may be a better one. These factors and associated risks should be considered as part of prioritising equipment for replacement alongside initiatives for service development or to meet new clinical and regulatory requirements (see Chapter 5).

At each stage of procurement, steps must be taken to ensure potential suppliers comply with corporate and financial law and best practice. A check must be done to confirm whether the device they intend to supply complies with legislation and standards for its intended use, particularly where new or novel applications are intended. The purchasing organisation must ensure conformance with local regulations and laws relating specifically to procurement, such as requirements for tendering, and to other governance aspects including its own financial rules.

A clear, written purchase specification is essential. This sets out the functions a medical device is expected to perform and any regulatory, technical and safety standards it must comply with. It is the basis for evaluating and testing different devices and for clarifying contractual arrangements. Guidance and assistance on writing specifications and carrying out evaluations should, where available, be sought from national government, advisory and professional bodies and colleagues, as learning from the experiences of others is invaluable in helping to get a device which delivers what is wanted.

Finally, the organisation must seek to obtain good value for money in procurement, not only with respect to purchase price but also when negotiating prices and arrangements for training, consumables, spares, maintenance and management. Contractual arrangements should be checked carefully to see how liability for device failure or damage is to be covered and what might be apportioned to the organisation if certain conditions are not met.

### 3.5.2 Installation, Acceptance and Commissioning

Medical equipment needs to be properly installed and set up prior to going into clinical use. Responsibility for installation needs to be agreed between the manufacturer and end user prior to purchase, in consideration of manufacturer recommendations and user capability. Acceptance tests should show that purchased device functions correctly and complies with performance characteristics set out in the purchase specification, including meeting all safety standards. Specialist test equipment and expert advice may be needed to carry out required tests, as even large and expensive items of equipment may not meet promised performance and safety standards after installation. An organisation's acceptance procedures should also include logging equipment onto an inventory for subsequent management, such as setting up maintenance arrangements and tracking its ownership and location in case of a subsequent equipment recall.

Where medical devices are connected to IT networks, potential risks include data loss and incorrect equipment operation from network and interfacing problems, and from breaches of data security. Careful planning helps to reduce these risks, and responsibilities must be assigned for each element of the network and device connection to manage both the infrastructure and its operation (see Appendix A).

### 3.5.3 Risks during Equipment Operation

#### 3.5.3.1 Operating Risks

Published and anecdotal data suggest the majority of serious medical equipment-related events stem from one of the three causes: user error (incorrect equipment set up or the wrong type of equipment used), equipment unavailable and faulty equipment. These themes reoccur regularly in incident reports from individual organisations. *User error* can be difficult to identify and is even more difficult to reduce. A notable number of incidents where equipment is labelled faulty end up with no problem being found, and it is likely that many of these events were not true equipment faults but were caused by user error. User error is reduced by improved training and working practices and also more fundamentally by better designed medical equipment. *Equipment availability* relies on good housekeeping and forward planning, to make sure it can be decontaminated and set up in time and that adequate consumables are ready. Judicious use of spare systems can save an organisation cancelling patients if a problem occurs, but this is impractical where devices are expensive and timely equipment replacement based on records of past reliability will also enhance availability. *Faulty equipment* is less likely to occur where effective maintenance is carried out. Monitoring the condition of equipment and analysing causes of breakdown enable a maintenance service to pinpoint areas of potential failure, with targeted inspections to maintain or replace critical parts before they fail. Although medical equipment is often of sophisticated and advanced design, the most common operating failures and risks encountered in practice arise from simple mechanical and electrical causes such as component wear, loose fasteners and damage to leads and connectors. Equipment failure or incorrect operation can result in injury to patients and staff and also damage other equipment in a way that may be unseen and create future problems. Clinical engineers can contribute significantly to reducing operational risks by training users and working with them to monitor and inspect for early signs of failure. Jacobson and Murray [5] give many examples of device-related incidents, highlighting injuries sustained, and suggesting likely causes and remedial actions.

### 3.5.3.2 Health and Safety Risks

Multiple health and safety risks arise when working with medical equipment. Some are tightly regulated, such as the use of ionising radiation and laser light, whilst others are far less widely appreciated including a potentially fatal risk from inadvertent tiny electrical shocks to the heart. A risk assessment should be carried out for each item of medical equipment and for any non-medical items that are integral to the way medical equipment is used, so that appropriate measures can be put in place to minimise risks to staff and patients. However, awareness of hazards is not in itself enough. Even when risks are well known and how to avoid incidents is well understood, it is all too easy to overlook a key safety precaution. Constant vigilance is required to prevent familiarity turning compliance into automatic behaviour [15] or a tolerance of shortcuts, and this is one reason for the disciplined and routine use of checklists in operating theatres and aircraft cockpits. Medical equipment hazards can fall into a number of areas such as radiation, electric shock, mechanical injury, chemical contamination, biological contamination, burns and fire. We outline these later as common examples of factors that need to be taken into account during risk assessments with medical devices.

### 3.5.3.3 Radiation Safety

Inadvertent exposure of staff, relatives or members of the public to ionising radiation occurs in spite of stringent regulation and controls being in place. The wrong patient may be x-rayed or an incorrect dose administered in radiotherapy, but serious incidents are relatively rare, as are harmful doses to relatives or staff or radiation leaks to the environment. However, constant vigilance is needed to keep the error rate low and make sure equipment generating, using or measuring radiation is calibrated and used correctly. Radiation shielding in health facilities must protect patients, staff and the public, and diagnostic and therapeutic procedures must minimise unnecessary radiation exposure. This is a complex and difficult area that requires specialist medical physics advice and support to manage.

### 3.5.3.4 Non-Ionising Radiation

Potentially hazardous levels of laser, infrared and ultraviolet light, radio waves and intense light sources are used in healthcare, for example, in physiotherapy and phototherapy. They may cause serious burns or damage to sensory organs such as the eye in both patients and staff. In the United Kingdom, the use of moderate- to high-powered lasers requires suitable safety precautions for staff under the European Union (EU) Artificial Optical Radiation Directive [16] and associated UK legislation [17–19] including the provision of formal risk assessments, user training, systems of work, warning systems, interlocks and suitable protective equipment.

### 3.5.3.5 *Electrical Safety*

Every healthcare professional must take extra precautions with electrical items in the clinical environment. There is a duty of care to protect patients, staff and users from electrical hazards associated with the use of medical devices and to protect staff from the adverse effects of electricity in the workplace. Staff include technical staff dismantling and repairing equipment.

Electrical equipment in a medical environment requires more stringent care than domestic and industrial equipment, as there can be a direct electrical connection or current pathway to the patient through items such as ECG and skin electrodes. In the case of invasive catheters or intraoperative probes, these connections can be internal, and severe injury or death can be caused by levels of electrical energy well below what can normally be felt. An unconscious or sedated patient is unable to react to sudden shocks or may develop electrical burns from causes such as poorly conducting diathermy electrodes. The clinical environment contains conducting fluids such as saline and blood which can inadvertently carry electrical currents to patients and staff. Electrical equipment is vulnerable to damage from movement and handling by users who are not aware of the safety precautions needed and possible consequences of their actions, and equipment used in critical situations such as operating theatres and critical care units is especially vulnerable, with regular staff training and equipment inspection being vital for equipment on which treatment or diagnosis depends. Electrical leads are a weak point with frequent damage to mobile equipment cables and plugs, and extension leads in the medical environment must be specifically approved and limited in their use. A less obvious risk arises when connecting medical to non-medical devices such as computers, for enhanced monitoring or research purposes, unless the medical equipment is specifically constructed for this purpose and manufacturer instructions are followed.

Electrical hazards to staff and patients range from mild discomfort to death by ventricular fibrillation or respiratory muscle paralysis. Injuries can arise directly from shock or burns, or secondary shock effects such as falls and sudden muscular reflexes. Electrical faults are also a common cause of fires, principally due to overheating cables, connectors or equipment. Electrical sparks can also ignite vapours or materials. Electrical fire can be avoided by proper maintenance, by regular inspection, and by fitting appropriate fuses, but fortunately, electrical fires are rare in clinical areas, due to regular safety inspections and testing. The major causes of fire in NHS hospitals are electrical faults in the building infrastructure and deliberate arsons, rather than problems with medical equipment.

### 3.5.3.6 *Mechanical Hazards*

Injury can arise from tripping and dropping hazards, patient or staff collision with moving equipment or support equipment giving way due to damage or overloading. The safe working loads of items such as theatre tables and trolleys can be exceeded by a combination of heavy patients and added accessories. Using medical equipment to support, transport or store inappropriate equipment or objects can result in its tipping over, as can using wheelchairs incorrectly. It can be difficult to test mechanical items of equipment adequately, and it is often not undertaken correctly. It is not unknown for ceiling or wall-mounted lights or equipment to work loose over time and to then fall on patients and staff. Regular inspection is the only way to prevent this type of failure. Staff injury from repetitive operations with medical equipment is an occupational hazard, for example, in sonography, and needs to be addressed alongside compliance with more general manual handling regulations.

### 3.5.3.7 Chemical Contamination

Hazardous gases and vapours are used in anaesthesia (Figure 3.3) and can leak from the patient breathing circuit into the environment. Active scavenging may be needed to avoid exceeding safe exposure levels. Nitric oxide therapy produces toxic nitrogen dioxide gas and environmental levels may need monitoring. In the medical equipment maintenance workshop, hazardous chemicals are used for cleaning and soldering, and toxic or respiratory irritants can be released during simple machining processes. Suitable safety precautions should be taken.

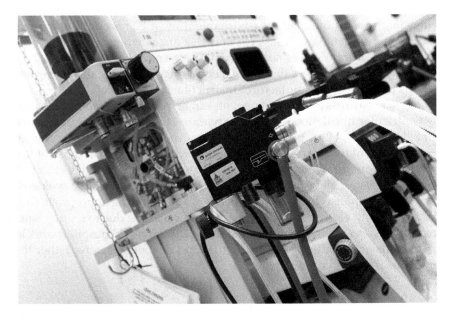

**FIGURE 3.3**
Anaesthetic machine and connections to patient breathing circuit. (Reproduced by permission of the Institute of Physics and Engineering in Medicine. Copyright IPEM.)

### 3.5.3.8 *Infection*

Poorly decontaminated equipment is an obvious infection risk. Although its effects are difficult to quantify, attention to equipment cleaning is one important factor in reducing cross-infection rates in healthcare facilities. Effective cleaning and disinfection of equipment plays a major part, alongside the appropriate use of disposable items.

The audit of policies and practice relating to single-use devices has been strong in the United Kingdom, particularly in relation to concerns regarding prion-based diseases such as Creutzfeldt–Jakob disease (CJD) which has led to some equipment being quarantined after use on one suspect case, with no ability to decontaminate it effectively after use on one suspect case.

The decontamination requirements of each type of medical equipment need to be checked carefully before purchase, to make sure an organisation can process and reuse it safely. The decontamination facilities available to a healthcare organisation will to some degree determine the amount of equipment it needs to deliver a service, as prolonged decontamination cycle and transport times, combined with a short sterile shelf life, can require more equipment circulating through the system. This is most critical for theatre and endoscopy instruments and where specialised or high-volume invasive procedures are carried out. Concern about prion transmission is leading to greater use of single-use or patient-specific medical devices for certain procedures, whereas decontamination requirements for most items of general medical equipment can be met using relatively straightforward techniques [20]. It is wise to note that approaches to reprocessing single use medical devices vary between different countries, ranging from declaring it illegal to supporting local reprocessing to meet clinical need, or even to legislation creating a national industry. Further information is available from the Association of Medical Device Reprocessors (AMDR).

### 3.5.3.9 *Heat Injury*

Burns can occur from physical contact with hot or cold surfaces. Prolonged contact with even moderately elevated temperatures (43°C or greater) can lead to severe burns [21] and can occur where equipment malfunctions or is incorrectly used, for example, where items are placed on unconscious patients. There has also been concern about whether the physical temperature of ultrasound probes can be high enough to cause injury but no actual incidents have been reported. Devices such as diathermy electrodes and operating theatre patient warmers need to be carefully used and checked for correct operation, particularly with neonatal patients where tolerance levels for skin injury are lower. Where techniques such as cryotherapy are used, cold burns can be caused inadvertently or through the incorrect routing of cooling pipes. Hypothermia is a more general patient risk that results from the environment, through overcooling and exposure in cold operating theatres, for example, and is not a specific risk for patient devices.

### 3.5.3.10 Electromagnetic Interference

Medical equipment such as pacemakers and sensitive diagnostic devices may be susceptible to electromagnetic interference, of which surgical diathermy and MRI systems are two well-known sources. Other potential sources interference in the medical environment include hand-held radios used by portering and security staff, wireless telemetry for physiological signals, and mobile phones used by staff and patients. There is evidence to indicate that high-power wireless devices interfere with life-critical equipment [22] and some older types of syringe and infusion drivers have also been shown to malfunction when very close to mobile phones. Evidence from the United Kingdom suggests that early radio frequency asset tagging readers could adversely affect operation of some life-critical devices such as pacemakers if very close to them, in the same way as retail security systems [23], and also that the magnetic field from some MP3 player headphones can affect pacemakers and implanted defibrillator/pacemakers [24]. However, the transient and varying nature of these signals makes it hard to identify them as a cause of specific incidents involving patient injury, even if suspected. An organisation has to strike a balance between the true risks and benefits of, for example, mobile phone use where as a rule, if equipment function is affected at all, it is from signals emitted at a distance of well under a metre. Newer equipment is less likely to be affected, due to more stringent immunity regulations [23]. The most serious potential interference problem is a subtle change to equipment settings, operation or function, whereby a specific function is altered or turned on or off, including delivery rates on infusion devices. Such an event could be life threatening if a syringe driver delivers a drug intended for infusion over several hours in a single bolus at its maximum delivery rate. Interference may also obscure physiological signals or give a misleading result. This is less serious if an operator is aware of the problem, so the fact that surgical diathermy causes serious distortion of monitoring signals is well understood. However, less obvious interference on, say, a 12-lead ECG may mislead a less experienced operator.

In summary, general telemetry and computer communications such as Bluetooth or cordless mice are not likely to interfere with medical equipment operation and neither are mobile phones, unless within a metre or less of older equipment. Devices giving the greatest cause for concern are two-way radios and diathermy equipment. Specific steps to minimise risks range from controlling where devices are used to specialist installation of cables and devices to shield them from interfering signals [25].

### 3.5.4 Maintenance

Regular and routine maintenance of equipment brings significant clinical and safety benefits. Managing maintenance effectively keeps costs down, increases equipment availability and helps ensure continuity of service provision. Regular user checks and calibration are needed for some diagnostic equipment to ensure results are accurate, and for the safe and effective operation of therapeutic equipment ranging from radiotherapy machines to syringe drivers. Equipment should be serviced at appropriate intervals and be repaired by trained staff, using appropriate spares. User maintenance includes regular quality control checks, cleaning, effective decontamination and visual inspection and has a key part to play in making sure equipment is reliable and safe, as does correct equipment storage and the use of appropriate consumables.

Failure to maintain equipment adequately results in more breakdowns and incidents. Lack of maintenance may also be picked up after a patient incident, even where equipment was not the primary cause, as investigations often uncover shortfalls in equipment management that can contribute to the risk of an organisation being found negligent. There is however a difference between simple failure to maintain an item and a situation where an organisation decides that adequate or appropriate maintenance is different to that recommended by the manufacturer. For example, replacement accessories such as pulse oximeter finger probes are available more cheaply from third-party suppliers than original manufacturers, and their use can represent a significant saving to a large organisation. In order to limit any liability, an organisation should be able to make a case that use of the alternative product is not deleterious to patient care or overall safety and is reasonable in the light of current knowledge. This may involve consultation with clinical engineering colleagues and a project to validate the use of alternative parts. Because of the potential savings, this is commonly done not only for peripherals but also for major components and consumable items such as batteries and infusion sets. What is important is that an organisation obtains expert advice and risk assesses any such decision, to avoid inappropriate use of items unsuitable for their intended purpose. This is another reason to work with professional colleagues, as liability is likely to be reduced where it can be shown that an approach is reasonable when compared to those taken by other organisations, in the light of their pooled experiences and outcomes. The same issues apply to the use of third-party maintenance services. An important principle is to remain sceptical of the quality of all parts and services received, as there are many examples of poor-quality parts and services inadvertently being supplied by the original manufacturer or by third parties.

Organisations cannot assume that purchasing equipment support from a manufacturer will guarantee a high-quality service, and performance monitoring of contracts needs to be carried out. Support is also likely to be limited in scope, and organisations may reap significant benefits from taking a more proactive approach than some manufacturers are able to provide. For example, preventative maintenance reduces the likelihood of breakdown and is usually carried out to the manufacturer's set schedule. The most proactive approach to reducing premature failure is to analyse the causes of equipment breakdown, to look for patterns of component or system failure.

There are well-developed statistical approaches for doing this, but reasonably large numbers of events are needed for the analysis to be effective, so they are best applied to common items in an organisation and indeed are used by manufacturers to improve their products and reduce servicing costs. Whilst large-scale statistical analysis may not be feasible for the majority of specialist equipment in an individual organisation, a lot can be achieved through careful observation and monitoring of breakdowns to identify common ways in which equipment is failing locally and to then plan ways to avoid premature failure. This can involve carrying out additional maintenance or replacing components in a planned way after an elapsed time or performance limit is reached. Manufacturers recognise the value of this approach and may incorporate simple visual wear indicators in mechanical systems or provide access to variables such as electrical motor drive currents that can indicate increased working loads and hence impending failure. It is however up to each organisation and its clinical engineering service to decide how best to make use of opportunities to reduce failures and costs.

### 3.5.5 Hazard and Incident Reporting and Management

Periodic quality control checks, carried out by equipment users or maintenance staff, have an important part to play in picking up potentially hazardous situations by identifying where an item is no longer performing to specification. This allows suitable adjustments, recalibrations or repairs to be made and is an integral part of regular monitoring. Newer equipment often has a range of inbuilt checks to make sure users are not operating it incorrectly alongside expert functions to diagnose errors and monitor operation. These checks help to keep equipment operating safely. As more functions and checks are incorporated, however, overall complexity increases, and the associated quality control and tests to check for correct operation become more difficult to do. Modern computer software, for example, is so complex that it is not possible to check every potential combination of environmental conditions and machine settings for inherent weaknesses and a fault in a programme embedded in a device may emerge at any time. Continued surveillance of medical equipment is therefore critical, particularly where it incorporates expert knowledge which modifies its function and is difficult to test.

When medical device problems do occur, it is essential to have a system for reporting incidents and near misses. This will include procedures for investigating incidents, taking preventative action and disseminating warnings, hazard notices, product recalls and general evidence-based safety advice across the organisation. Reporting incidents to national regulators and safety bodies is the responsibility of both equipment end users and manufacturers. Incident reporting takes place at two levels. The first is within the organisation, to bring risks and hazards to the notice of risk managers so that internal issues involving incorrect use, poor training and faulty procedures can be rectified quickly. The second level is external reporting to regulators or other bodies that can pick up common threads across different incidents when looking at equipment design, operation or use. Regulators have the power to withdraw manufacturers' products or make them act to address hazards and disseminate warnings and advice to users. The relevant regulatory body in the United Kingdom is the MHRA and in the United States is the Food and Drug Administration (FDA).

It is an apparent paradox that organisations reporting more incidents actually reduce the risk of serious ones. This is because causes of near misses are similar to those of major incidents and investigation results in changes in procedures and practices that eliminate both [26]. The concept of the *Heinrich ratio* [27] was first published in 1931, based on data from the insurance industry. It established that, on average, for every reported major injury there were 29 incidents of minor injury and 300 near misses, so the identification of hazards via near misses is far more likely and far less damaging than waiting for major events. In the absence of encouragement to report all incidents, organisations tend to report only the relatively small number of serious occurrences. At the inception of widespread near-miss recording in an industry, the number of reports initially goes up but in the long-term there are falls in the number and proportion of major incidents. A good example of this is the international aviation industry which started an intensive safety programme in 1989. In 1994, 4500 reports were filed of which 2.7% were high risk. By 1999, 8000 reports were filed, of which 0.2% were high risk. Subsequent mathematical analysis has shown that the Heinrich ratio can vary substantially within and between different processes but that the conceptual principle behind it holds in most cases [28].

The organisation will require a single traceable line for disseminating hazard and advisory notices and product recalls, together with feedback from users to confirm necessary actions have been initiated and completed. When a device is subject to manufacturer recall for modification or replacement, the equipment management service will need to check whether the organisation has that equipment and must then retrieve, dispatch and replace affected devices or liaise with manufacturers to arrange for this to be done. It must also monitor progress and sign off when the project is completed. This is where keeping up-to-date records of equipment ownership and location on the organisation's medical equipment database proves its worth, underlying the importance of comprehensive acceptance procedures and updated inventories able to verify the existence and location of devices in the organisation. In practice, it can take significant time to complete a device recall even where good tracking procedures are in place.

### 3.5.6 Modification, Clinical Trials, Research and Off-Label Use

Increasing levels of medical equipment legislation, coupled with risk aversion by organisations and insurers, can cause difficulties when proposing to build or modify equipment in-house. The benefits of doing so apply largely to organisations carrying out research or innovative procedures and can be considerable, leading to improvements in clinical care and commercial returns from licensing or spin-out companies. To carry out such work, organisations must have the necessary skills, knowledge and commercial contacts and risk management procedures that allow them to modify and produce devices in a way that meets the demands of regulatory bodies. Such procedures need to be set up formally and must be subject to internal and external audit. Examples of activities in this area include the following:

*Device modification*: One example is obtaining agreement from a manufacturer to fit an alternative power supply transformer and make other modifications allowing an item to be used in a country with different power supply voltages. Where modifications are made in agreement with the manufacturer, they will retain liability for consequent faults, but if an organisation modifies equipment independently, it becomes the manufacturer and assumes liability. This can include a step as simple as using an incorrect spare part, such as a replacement screw that is too long and which then contacts a live cable.

*Clinical investigation*: This includes construction of a novel device, for example, to record electrical signals from the body during MRI procedures or the connection together of disparate systems not designed to be used in that way – such as connecting a computer to a medical device without a suitable output.

*Off-label use*: This refers to the use of a medical device for a purpose for which it has not been approved or was not intended, such as using a catheter for neonatal feeding in the absence of suitable CE-marked alternatives. Medical devices are manufactured and licensed with a specific clinical end use in mind, and to use a device off label for another purpose requires a risk assessment and discussion with regulators in the light of current practice.

Medical devices, unlike consumer devices, cannot be expected not to do harm, but their use represents a probable benefit and a patient, by giving informed consent, accepts this. However, many claims are brought on the basis that the patient was not warned sufficiently of the dangers they were facing, and using a device made in-house opens up an organisation to product liability or other legislation for damages. Major categories of potential liability include defective design or manufacture, faulty warnings and negligent surveillance. An organisation can reduce liability when carrying out modifications and manufacturing by

- Writing clear specifications for functions and outcomes
- Being able to track the source of parts, sub-assemblies and expert advice back to suppliers and advisers

- Supplying adequate instructions and warnings for safe use
- Using quality management and assurance procedures at all stages of production
- Designing out errors and potential problems through a thorough risk analysis
- Checking conformity with regulations and construction standards governing the product
- Taking steps similar to other manufacturers to address risk and liability issues
- Maintaining design and process records, especially regarding the current state of scientific and technical knowledge, for at least 10 years after production
- Keeping detailed records of components, inspections, sale and delivery
- Seeking to transfer liability for modifications back to the manufacturer by obtaining their permission or approval
- Having adequate insurance cover, including product liability insurance
- Checking suppliers have liability cover
- Using the original manufacturer's subcontractors where possible
- Continuing to monitor items once issued, to pick up any failures or other problems

Medical equipment is often connected to computers, including for research purposes. If a computer was not supplied or approved by the device manufacturer, any connection presents a potential electrical safety and patient data security risk, so additional precautions may be needed when using them for equipment control or research (see Chapter 12). Also once identifiable patient data have been transferred to a computer, it becomes subject to data security constraints. Recording and reporting of test results and the storage of data for research must be carried out under strict conditions of security, monitored in the UK NHS by nominated *Caldicott guardians* for each organisation. Data security is increased by anonymisation, encryption, password protection and other access control measures. The overriding principle is that patient identifiable data should only be communicated to or be accessible by those who need to use it, and they should only see as much of the available data as is necessary for them to perform their function within the organisation.

### 3.5.7 Condemning and Disposal

Condemning and disposal must be carried out in line with the organisation's financial and corporate practice. There are usually transparent and auditable procedures to follow when disposing of equipment, to avoid the risks of fraudulent resale and improper reuse of unsafe or condemned equipment for clinical or research purposes. Chapter 12 goes into more detail on equipment disposal and related risks, highlighting relevant laws and regulations including the need for thorough decontamination and the removal of any patient data. Specialist advice is necessary for some devices, such as those containing radioactive sources.

## 3.6 Liability and Indemnity: When Risk Becomes Reality

### 3.6.1 Liability

Liability is unavoidable for a healthcare organisation. In spite of rigorous risk management, accidents and mistakes still happen with tragic or expensive consequences. Once an event happens, any consequences for patients are best met by a rapid, sympathetic response and appropriate compensation. The level of clinical claims is such that hospitals will insure against financial loss either directly with an insurance company or through some type of pooled scheme. To limit the likelihood and extent of any payout, insurance schemes create incentives for organisations to follow best practice and may inspect them regularly. Similarly, an organisation must ensure that its suppliers and contractors are appropriately insured to be able to compensate it for any damage caused by their goods and services including medical devices which have been purchased, are on loan or are undergoing clinical trials.

### 3.6.2 Indemnity

Indemnity is defined as *'Security from damage or loss'*. Specialist schemes or insurers provide cover for the costs and damages awarded through civil proceedings to individuals seeking compensation for medical injury claims. In large nationally run healthcare organisations such as the UK NHS, pooled risk-sharing schemes and the certainty that there will be an ongoing level of claims mean that standard insurance for individual organisations is not likely to be cost-effective unless they specialise in low-risk procedures or need specific top-up insurance for particular activities or less common risks. This should include cover for any external contracts providing servicing, equipment management or consultancy to other organisations and for any manufacturing activities. It can be expensive to add on additional functions but good practice and audit can keep premiums down. Since an individual healthcare practitioner may also be sued, professionals require their own indemnity cover if carrying out private activities such as consultancy work on their own account rather than through their employing organisation.

It is up to the individual to establish what is and what is not covered by their organisation's insurance, as well as obtaining approval for any external activities. Many organisations hold a register of interests where such activities must be declared.

## 3.7 Legal Obligations of Healthcare Organisations

### 3.7.1 Introduction

Understanding the differences and relationships between legislation, regulations, standards and best practice puts requirements for medical device management into context and clarifies what is necessary and what is discretionary. *Statutory legislation* sets out what must (or must not) be done, either by specifying or proscribing actions or by defining individual and corporate duties and responsibilities. *Regulations* are specific implementations or embodiments of legislation that can be updated as the need arises without invoking the state legislature, through an identified role or body such as a health minister, health ministry or other competent authorities. *Standards* contain particular guidance or specifications for the construction and performance of particular types or classes of device and for management processes, and are agreed by national and international standards bodies (see Chapter 13). *Best practice* is advisory and is exemplified by professional guidelines and advice from government bodies on applying legislation and regulation.

Although only legislative requirements and subsequent regulations are mandatory, compliance with legislation often implicitly requires compliance with relevant standards. For example, the essential requirements of the EU Medical Devices Directives include compliance with appropriate international standards on medical devices and electrical safety. Any defence in law may well depend on showing that best practice guidelines were followed, as well as relevant legislation. The clinical use of medical devices is regulated through national licensing or authorisation schemes allowing healthcare facilities to operate and set up standards and inspection bodies. Relevant UK legislation is the Health and Social Care Act [29] with the inspection body being the Care Quality Commission. In the United States, each state approves its own system and operates under State and Federal law.

Legislation is not always easy to interpret, particularly when complex issues are involved. This leads to uncertainty, and ambiguities are usually resolved through decisions in the courts or exceptionally by redrafting legislation. In the absence of a formal legal opinion, the best way to reduce an organisation's exposure to legal action is to consult with regulatory bodies and professional peers and seek to act in a way consistent with common practice and the spirit and language of the legislation.

### 3.7.2 Legislation

Each country has its own legal systems to govern how individuals and organisations should behave. These systems set out penalties for misbehaviour or systems of redress for wrongs carried out. As an example, English law is divided into two areas: Statutory law is produced by Parliament, and common law based on previous decisions and the outcome of legal cases. Statutory law is produced by legislative bodies, such as the Westminster and Scottish Parliaments in the United Kingdom. They are stated in written form as Acts of Parliament and can both proscribe and prescribe actions and responsibilities intended to regulate specific activities. Criminal acts or non-compliance with mandatory requirements can result in prosecution by the state (the Crown in the United Kingdom) with subsequent imposition of fines and/or imprisonment. Criminal liability arises mainly from statute law although, for example, manslaughter is an offence under common law. Civil cases are brought by aggrieved individuals or organisations seeking redress against other individuals or organisations alleged to have breached contracts or practice as defined by statutory or common law. Unlike the criminal court, where the accused is either found guilty or innocent, both parties may be found to have contributed to a loss or injury, and hence only a proportion of compensation may be paid and legal costs may be allocated in whole or in part to either side.

Legislation that concerns medical devices is of three kinds. The first comprises laws concerned with the production, use and surveillance of medical devices, such as European medical device law. The second is health and safety legislation, which aims to protect workers and members of the public from hazards. It applies to any commercial or public environment, including those in which medical devices are produced and used. The third kind is consumer protection law, which seeks to protect a purchaser against harm from faulty products by providing a mechanism for seeking direct compensation from the manufacturer or supplier. This mechanism creates financial penalties to deter manufacturers from producing faulty or unsafe goods but cannot of itself prevent the sale of such products.

### 3.7.3 Medical Device Law and Regulations

Medical device law principally covers the responsibility of manufacturers and suppliers. It seeks to ensure that the design, production, testing, documentation, marketing and ongoing surveillance of medical devices meet local requirements, regulations and standards. Although it is the responsibility of a supplier to ensure compliance when placing a device on the market, the use of a non-compliant device can present a risk to the user, and so users need to know what local requirements are in order to specify and check device compliance. The user also needs to take responsibility for any device used *off label*, where it is operated in a way which does not comply with the original device approval, for example, using it in a modified form or for a clinical purpose not intended by the manufacturer.

The pattern of medical device legislation varies from country to country. Many countries operate a similar system of device law to Europe, such as China, and we will use the European system as our primary example in this chapter. In Europe, medical devices legislation comprises the Medical Devices Directives, which are put into operation in each European country by national legislation and overseen by a competent body, such as the MHRA in the United Kingdom, supported by relevant technical standards for medical electrical equipment which are accepted as those set out by the International Electrotechnical Commission (IEC). In the United States, the relevant national body is the FDA, whilst some production standards are devised by the National Electrical Manufacturers' Association (NEMA). These and other legislative and regulatory bodies worldwide are introduced in Chapter 13. In order to illustrate issues raised by the regulation, the following paragraphs look in more detail at practice in Europe.

The European Medical Device Directives [30] regulate objects placed on the market (excluding drugs) within the EU intended for use as medical devices. They were introduced to facilitate a single European market. They allow medical devices designed and manufactured to harmonised standards to cross boundaries between national states without additional regulation. Compliance in one EU member state means compliance in all. Three original directives are due to be reduced to two regulations under proposals adopted by the European Commission in September 2012 and submitted to the European Parliament and Council for approval. These regulations cover medical devices and in vitro diagnostics but retain much of the original directives, and their basic concepts.

### 3.7.3.1 Conformity

European regulations effectively demand manufacturers use a quality system to cover medical device production and design, with components, and management and production processes meeting relevant standards (harmonised across member countries) including labelling, the provision of documentation and post-marketing surveillance (including incident reporting). Devices are classified, for the purposes of regulation, as conforming to a particular class depending upon the severity of medical risk they pose. This risk classification is also relevant to the way in which medical devices are managed in organisations. Classification ranges from Class I for low-risk devices, for example, wheel chairs and electrodes, to Class III for the highest-risk devices such as cardiac pacemakers and other implantable items intended for critical applications. Other regulatory systems use similar staged classification systems.

### 3.7.3.2  CE Marking and Identification

The visible sign of conformity to EU standards is the CE mark, which should declare which standards a device meets. The CE marking system however applies across a wide range of consumer products, which can be a source of ambiguity. For example, a blood pressure monitoring device may be CE marked as conforming to a general electrical product standard for recreational use, enabling it to be purchased for self-monitoring by individuals, but not be approved for use in a medical context to support decisions on clinical care. For this latter purpose, the device would need to be CE marked as conforming to the relevant medical device regulations and associated standards. Clinical users should check with the clinical engineering service where they are uncertain. The proposed EU medical device regulations include permissive legislation to allow states to establish a record of individual medical devices using a universal device identifier, which requires manufacturers to uniquely identify every device they produce. Similar systems are being proposed or set up worldwide, for example, in the United States. The EU proposal is that healthcare organisations keep an electronic record identifying all devices under their control, report individual device details when logging clinical incidents, and keep a record of which device was used in *high-risk* procedures or with *high-risk* patients, although the term *high risk* is not yet defined.

### 3.7.3.3  Placing on the Market

*Placing on the market* is a key concept. It includes making a device available for sale *and* any other method of supply, such as loan, lease, hire or donation. A medical device may not be placed on the market within the EU unless it is CE marked as conforming to the relevant medical device standards and is effective for its intended purpose. This measure is intended to ensure all medical devices are adequately evaluated and manufactured to agreed safety standards. One consequence is to prevent a healthcare or academic organisation from producing or modifying equipment for research or special treatments and supplying it to another organisation, unless that device is CE marked or in use as part of a registered device trial. This controls the diffusion of new technologies in the interests of patient safety, at the cost of proving device conformity, which can be complex and expensive.

### 3.7.3.4  Exemptions

Many medical devices do not carry a CE mark because they have not been put onto the market under the formal definition. These are mainly devices produced or fitted individually for named patients, following approved standards or modified to manufacturer specifications, such as dentures and individual wheelchair seating moulds. A healthcare organisation may also develop and produce custom one-off devices to meet special clinical needs for named patients or create a medical device solely for use on its own patients, and neither of these involves placing on the market. However, this does not relieve hospitals of their other obligations to ensure equipment is manufactured to appropriate standards and is safe to use. The essential requirements set out in medical devices regulations are categorised either as general requirements to manage risk and maximise safety or as specific requirements for standards of design and construction. Both need to be considered carefully when considering what will prove the safety of a device or when seeking CE marking for an in-house manufactured item.

A common misconception about CE marking is that it is illegal to use a device that is not CE marked. The law, however, regulates *'placing devices on the market for an intended use'*. A clinician who uses a medical, or any other, device for a novel clinical purpose is entitled to do so but must accept responsibility for managing associated risks, in some cases becoming the legal manufacturer of the device. They would be wise to document how they have addressed all risks and followed best practice, in case of any claim arising. A complementary myth is that CE marking on any device makes it implicitly suitable for medical use. However, it is the manufacturer's statement of conformance that shows whether or not equipment is approved for use as a medical device.

Notwithstanding these exceptions, nearly all medical equipment acquired by healthcare organisations in Europe will carry the CE mark. Each organisation will need to establish that it does so before considering purchase. In the United Kingdom, compliance with CE marking and standards is usually verified by requiring the manufacturer to complete a pre-purchase questionnaire (PPQ) that also contains a wide range of questions concerning product support, decontamination and maintenance requirements, future upgrades and expected product lifetime. This is usually scrutinised by the clinical engineering service, which will follow up any ambiguities.

### 3.7.4 Consumer Protection Law

Faulty products can cause injury, and compensation is available through consumer protection legislation. This class of law is relevant to medical device management because it provides redress for individuals against suppliers of goods, including healthcare providers and their suppliers. In the United Kingdom, the *Consumer Protection Act 1987* [31] implements the European Community Product Liability Directive, under which unlimited damages may be payable for direct personal injury. Anyone injured by a defective product can take action against producers, processors, assemblers, packagers, modifiers, own-branders and importers within certain time constraints. The Act provides the same rights to anyone injured by a defective product, whether or not the product was sold to them. It covers consumers, not businesses, so redress for damage to property such as a consequential fire caused by faulty goods must be sought through other civil law channels such as negligence or breach of contract. Liability under the Act is joint and several, and a plaintiff may sue all defendants. It is not possible to exclude liability under the Act by means of any contract term or other provision.

The Act covers all consumer goods and goods used at a place of work including components and raw materials. A defective product is one where, *'the safety of the product is not such as persons generally are entitled to expect'*. This includes defective design, defective manufacture, inadequate warnings and negligent surveillance, when a manufacturer does not properly warn consumers about a subsequently discovered lack of safety. When deciding whether a product is defective, a court will also take into account what it might reasonably be expected to be used for and any instructions or warnings that are supplied with it. Also a defendant's liability could be reduced if a plaintiff contributed to his or her injuries by careless behaviour.

Specialist legal advice is essential when considering whether to proceed with legal action over which route to follow, and also for organisations seeking to limit their legal liabilities. The ultimate determinants of liability are the decisions of the courts in each jurisdiction, and until a significant body of case law is built up, major uncertainties will remain.

### 3.7.5 Health and Safety Legislation

In addition to complying with specific medical device and consumer protection legislation, organisations have a duty to staff, patients, contractors and the public under health and safety legislation. All those concerned professionally with managing medical equipment need to be aware of these and other areas of legislation and of the need to seek specialist advice when carrying out activities with dangerous substances or undertaking work in areas that may be covered by specific regulations. In the UK, health and safety legislation is prosecuted in the criminal courts, in most cases either by health and safety executive inspectors or by local authority environmental health officers. Successful action can be taken in the civil courts under common law by injured staff, resulting in financial penalty or compensation. The concept of *duty of care* is important, and if an employer fails to take reasonable care to protect an employee from a foreseeable injury, damages can be awarded against him/her if found to have neglected this duty. An employer can be simultaneously sued by an injured employee for damages and be prosecuted under criminal law for the same injury. Most health and safety prosecutions in the UK hospital sector have been due to a failure to establish safe working systems and also for not reporting accidents [32]. The Health and Safety Executive in the United Kingdom may issue improvement and prohibition notices, and individuals may be prosecuted individually and fined or imprisoned for wilful or reckless disregard of health and safety requirements.

Health and safety legislation covers aspects of the use, care and disposal of medical equipment in the workplace. Each country's legal system will contain some legislation relevant to this, and the United Kingdom has a well-developed set of health and safety laws. Those most relevant to equipment management and maintenance are outlined later and some are discussed in more detail in later chapters. Helpful information on the interpretation and application of a number of these regulations is available from the UK Health and Safety Executive [33].

The *Health and Safety at Work Act (1974)* [17] is the primary legislation covering occupational health and safety in the United Kingdom. The Health and Safety Executive is responsible for enforcing this Act and related Acts and Statutory Instruments. The Act aims principally to

- Secure the health, safety and welfare of persons at work
- Protect others against risks to health or safety in connection with the activities of persons at work
- Control the acquisition, possession and use of dangerous substances
- Control certain emissions into the atmosphere

The related *Management of Health & Safety at Work Regulations (1999)* [18] require every employer assess their activities for risks to employees and others and then take steps either to remove or ameliorate those risks, in most cases as far as is reasonably practicable.

In equipment management, provisions of the *Health and Safety at Work Act* directly affect the ways in which maintenance and servicing are carried out, including the need for regular inspection. These duties are clarified by *The Provision and Use of Work Equipment Regulations 1998 (PUWER)* [34], which requires that risks from equipment used at work are prevented or controlled. Equipment must be safe for use, be maintained in a safe condition, be provided with appropriate safety information and markings and protective devices, and be used only by people who have received adequate training. These requirements apply both to medical equipment and to test and measuring equipment used within a clinical engineering service.

The *Manual Handling Operations Regulations (1992)* [35] and subsequent amendments aim to reduce the risk of injury from lifting, carrying and manipulating goods and materials. They require risk assessments of operations to be carried out, coupled with means of removing or reducing any hazards identified. The *Regulations* establish a hierarchy of control measures, starting with avoiding hazardous manual handling operations so far as is reasonably practicable. If this is not possible, the hazard might be reduced by redesigning the task in order to avoid moving the load or by using mechanical devices such as lifting equipment. Specific regulations in the United Kingdom, the *Lifting Operations and Lifting Equipment Regulations* (1998) [36], apply to the operation, care and inspection of lifting devices which include items such as patient hoists. This is a major concern for a healthcare institution dealing with incapacitated patients and a serious potential risk to the health of staff. The employer is required to carry out a risk assessment for all such operations, to keep records of assessment and to provide suitable equipment and improved working procedures as needed. Clinical engineering services will be involved in delivering or commissioning regular servicing and checking of patient hoists.

To achieve compliance with *the **Electricity at Work** Regulations (1989)* [37], an organisation requires proof that its electrical systems are safe. This involves amongst other things the proper inspection and testing of systems and appliances by competent people and the creation and maintenance of suitable records. This applies to the wiring infrastructure, fixed installations which are permanently connected to the electricity supply, such as x-ray sets, and also to portable devices. Specific guidance was produced in the United Kingdom by the MHRA for fixed medical electrical installations, the *Medical Electrical Installation Guidance Notes (MEIGaN)* [38], and if available it should be consulted along with the other standards and guidance it refers to. Compliance with the Regulations is enforced through inspection by local authorities and fines for failure to comply. The recommended advisory interval for inspection of fixed wiring in hospitals is every 5 years. Portable devices need to be visually inspected and tested by a competent person at appropriate intervals, according to a risk assessment for each type of equipment. There is no recommended interval but many hospitals adopt a period of 1 year between tests. The frequency of electrical inspection and testing is based ideally on a risk assessment for each type of device, taking account of the particular circumstances in which it is used. Such an assessment can form part of a wider risk assessment covering planned maintenance. The most common cause of electrical safety failure associated with portable devices is accidental damage to mains leads, plugs and connectors.

*Environmental protection* law includes the Waste Electrical and Electronic Equipment Directive (WEEE Directive) [39]. It aims to reduce the amount of electrical and electronic equipment being disposed of inappropriately by encouraging its reuse, recycling and recovery, and also aims to improve the environmental performance of businesses that manufacture or supply electrical and electronic equipment or process it for scrap or recycling. It is relevant to the scrapping and resale of electronic and electrical equipment by healthcare organisations. Chapter 12 provides more information on this and other regulations restricting the use of toxic metals and some organic chemicals in new medical equipment. This also affects the sourcing of components and spares for older equipment, and the in-house construction and maintenance of medical devices.

*Control of infection* in the United Kingdom comes under the *Health Act 2006* [40] and other regulations which set out measures to help NHS bodies prevent and control healthcare associated infections, including the need to properly decontaminate medical equipment before use on patients. Ineffective cleaning and decontamination is a risk to maintenance staff and is controlled under general health and safety legislation. Failure to observe suitable measures may result in an inspector demanding improvement action. Suitable decontamination methods must be identified and put in place before equipment is reused, and this can be difficult to achieve for specialist items requiring specialised methods of disinfection or sterilisation. Also the effectiveness of decontamination systems needs to be monitored, with a requirement to keep records of regular testing. Procedures where there is a high risk of CJD contamination on neurosurgical and other instruments require additional controls, including quarantining and potentially destroying equipment that might be contaminated. The provision of sterilisation services is outside the scope of this book but it is an area with its own specialist regulations and controls, in which some clinical engineering departments become involved.

The International Commission on **Radiological Protection** (ICRP) makes recommendations on exposure limits and other measures for protecting workers and the public from ionising radiation [41]. These measures are interpreted by national governments and agencies and implemented through national legislation and regulation [42]. The UK regulations that enact EU ionising radiation protection directives such as the Euratom *Basic Safety Standards (BSS) Directive* [43] include the UK *Ionising Radiations Regulations (1999)* [44] which sets out health and safety requirements for persons working with ionising radiation and radioactive substances, and the public. They are enforced and advised on by the UK Health and Safety Executive. Employer duties under these regulations include devising and implementing a radiation safety policy, appointing competent radiation protection advisors, undertaking prior risk assessments for procedures involving ionising radiation, devising safe systems of work, designating controlled areas and producing local rules for work within these areas, and the appointment of radiation protection supervisors to ensure local rules are followed. Clinical engineering staff who check and maintain items such as x-ray sets, CT scanners and linear accelerators and their accessories, must know and operate under these local rules and agreed systems of work.

*The* **Radioactive Substances** *Act* 1993 [45] is intended to control radioactive substances, providing security and traceability from cradle to grave. It is enforced by the UK Environment Agency (EA) and requires that organisations register prior to acquiring and using radioactive materials and mobile radioactive apparatus. Under the Act the EA authorises organisations to accumulate and dispense radioactive materials, and also authorises disposal of radioactive waste under the *Environmental Permitting Regulations 2010* [46]. The Act specifies the need for qualified expert(s) with relevant training to be employed, to ensure all requirements under the Act are met. Suitable procedures must be in place to trace, identify and account for all radioactive materials and thus enable any losses to be discovered. Requirements for record keeping are extensive and the organisation is responsible for complying with regulations regarding the storage, transport and shipment of radioactive material, and for monitoring and leak testing during use, safe storage and safe movement within the organisation. Clinical engineers may encounter instrumentation containing radioactive sources, for example, some types of laboratory or imaging equipment and must take expert advice before attempting to maintain, repair or dispose of them.

*Other relevant UK legislation relevant to radiation use* in the medical field includes the *Ionising Radiation Medical Exposures Regulations (2000)* (IRMER) [47] which regulate systems for exposing patients to radiation, including requirements for recording patient exposure during procedures and the quality assurance of equipment. The *High Activity Sealed Sources Regulations* [48] set out arrangements for the security of powerful radiation sources and impact on security and facilities staff.

*Wider control of hazards in the UK* is implemented through more general legislation including the *Control of Substances Hazardous to Health (COSHH) Regulations 2002* [49,50] and *Carriage of Dangerous Goods and Use of Transportable Pressure Equipment Regulations 2009* [51]. Their subsequent amendments and other specific regulations limit how inherently risky items can be used and transported. Clinical engineers should know about these general risks and how they apply to any manufacturing and maintenance activities or to the clinical services they provide.

*Reporting requirements* under health and safety legislation usually specify that serious incidents be raised with a Notified Body. More than one body may be involved where areas of legislation overlap. Incidents in the UK where medical devices cause injury are reported to the MHRA as notified body under EU medical device regulations, and additionally to the Health and Safety Executive (HSE) for equipment failure or significant staff injury under the *Reporting of Injuries, Diseases and Dangerous Occurrence Regulations* (RIDDOR) [52]. The latter makes occupational diseases and injuries to workers above a certain threshold legally notifiable to the HSE from whatever cause, including dangerous equipment. Such incidents include mechanical injuries sustained from commonplace items not immediately associated with risk. For example, from 2001 to 2009 a total of 21 deaths and numerous injuries involving bed rails [53] were reported under RIDDOR. Reporting also includes incidents involving nuclear materials and radiation which in the United Kingdom may be reported to the EA, Health and Safety Executive and the Care Quality Commission depending on the type of incident. Specialist advice on reporting should be sought from the radiation protection adviser of the organisation involved.

## 3.8 Summary

In this chapter, we have considered the risks healthcare organisations face through operating and supporting medical equipment. We have outlined nature and occurrence of these risks, and strategies to eliminate or reduce them. We have introduced some of the obligations a healthcare organisation has to address risk, whether to provide a safe and effective clinical service, comply with statute or provide a defence under civil law. We have suggested that effective risk management makes a significant contribution to improving financial and operational efficiency and to maintaining the good reputation of the healthcare organisation. We categorised risk into clinical, health and safety and corporate areas, ranging from physical hazards such as infection, ionising radiation and electricity to human factors such as poor training and user error. The risk management process includes identifying risks, analysing their importance, devising methods of elimination or control and monitoring outcomes, and it applies to all stages of the equipment management lifecycle. We saw that if things do go wrong, indemnity can limit the financial consequences to the organisation, and good governance and effective quality systems may help to protect it against claims of negligence.

Finally, we outlined the legislative background relevant to medical device management under three main headings: medical device law, consumer protection law and health and safety law. Medical device law regulates what may be placed on the market, setting out how to achieve compliance through demonstrating safety and efficacy and following up safety in use. Consumer protection law channels how civil claims for damages are handled. Health and safety law seeks to protect staff, patients and public alike from multiple hazards. Compliance and familiarity with all three areas is a necessary duty for clinical and engineering staff alike, and appropriate specialist advice is essential if getting embroiled in any legal process.

## References

1. Brockton, J. et al. Estimates of medical-device associated adverse events from emergency departments. *Am. J. Prev. Med.*, 27(3), 246–253, 2004.
2. NHS Expert Group on Learning from Adverse Events in the NHA. *An Organisation with a Memory*. The Stationery Office, London, U.K., 2000.
3. NHS Litigation Authority. Report and Accounts 2012/13. London: The Stationery Office. HC527, July 2013. http://www.nhsla.com/AboutUs/Documents/NHS%20LA%20Annual%20Report%20and%20Accounts%202012-13.pdf (accessed September 09, 2013).
4. Bartup, B. Personal communication, 2013.
5. Jacobson, B. and Murray, A. *Medical Devices: Use and Safety*. Churchill Livingstone, London, U.K., 2007.
6. Gigerenzer, G. *Calculated Risks: How to Know When Numbers Deceive You*. Simon and Schuster, New York, 2002.
7. AS/NZS 31000:2009 *Risk Management – Principles and Guidelines*. Standards Australia/Standards New Zealand, Sydney/Wellington, 2009. http://sherq.org/31000.pdf.

8. Vincent, C., Neale, G. and Woloshynowich, M. Adverse incidents in British Hospitals: Preliminary retrospective record review. *BMJ*, 322, 517–519, 2001.
9. Abraham, N. Personal communication, 2010.
10. Sawyer, C. *Do It By Design: An Introduction to Human Factors in Medical Devices*. U.S. Food and Drug Administration Center for Devices and Radiological Health, 1996. http://www.fda.gov/MedicalDevices/DeviceRegulationandGuidance/GuidanceDocuments/ucm094957.htm (accessed August 26, 2013).
11. Wright, J. and Hill, P. *Clinical Governance*. Elsevier Science, Edinburgh, U.K., 2003.
12. Emslie, S., Knox, K. and Pickstone, M. (eds.). *Improving Patient Safety: Insights from American, Australian and British Healthcare*. ECRI Europe, Welwyn Garden City, U.K., 2002.
13. Scally, G. and Donaldson, L.J. Clinical governance and the drive for quality improvement in the new NHS in England. *British Medical Journal*, 4 July, 61–65, 1998.
14. Tennant, G. *SIX SIGMA: SPC and TQM in Manufacturing and Services*. Gower Publishing, Aldershot, U.K., 2001.
15. Toft, B. and Mascie-Taylor, H. Involuntary automaticity: A work-system induced risk to safe health care. *Health Serv. Manage. Res.*, 18(4), 211–216, 2005.
16. European Commission. Non-binding guide to good practice for implementing Directive 2006/25/EC. 2011, pp. 1–144. http://bookshop.europa.eu/is-bin/INTERSHOP.enfinity/WFS/EU-Bookshop-Site/en_GB/-/EUR/ViewPublication-Start?PublicationKey=KE3010384 (accessed August 26, 2013).
17. U.K. Parliament. Health and safety at work act 1974. http://www.legislation.gov.uk/ukpga/1974/37/contents (accessed August 26, 2013).
18. U.K. Statutory Instruments 1999 No. 3242. The management of health & safety at work regulations 1999. http://www.legislation.gov.uk/uksi/1999/3242/contents/made (accessed August 21, 2013).
19. U.K. Statutory Instruments 2010 No. 1140. The control of artificial optical radiation at work regulations 2010. http://www.legislation.gov.uk/uksi/2010/1140/contents/made (accessed August 21, 2013).
20. U.K. Department of Health. *Choice Framework for Local Policy and Procedures 01-01 – Management and Decontamination of Surgical Instruments (Medical Devices) Used in Acute Care*. U.K. Department of Health, London, U.K., 2012.
21. Greenhalgh, D. et al. Temperature threshold for burn injury: An oximeter safety study. *J. Burn Care Rehabil.*, 25(5), 411–415, 2004.
22. Barbaro, V. et al. Electromagnetic interference by GSM cellular phones and UHF radios with intensive-care and operating-room ventilators. *Biomed. Instrum. Technol.*, 34(5), 361–369, 2000.
23. Oduncu, H. *Use of RFID in Healthcare Settings and Electromagnetic Interference of RFID Devices with Medical Equipment: Review of Current Standards and Case Reports*. University of Glamorgan, South Wales, U.K., 2008.
24. Lee, S. et al. Clinically significant magnetic interference of implanted cardiac devices by portable headphones. *Heart Rhythm J.*, 6(10), 1432–1436, 2009.
25. U.K. MHRA. 2004. Mobile communications interference. http://www.mhra.gov.uk/Safetyinformation/Generalsafetyinformationandadvice/Technicalinformation/Mobilecommunicationsinterference/index.htm (accessed August 21, 2013).

26. NHS Expert Group on Learning from Adverse Events in the NHA. *An Organisation with a Memory*. The Stationery Office, London, U.K., 2000.
27. Heinrich, H. *Industrial Accident Prevention. A Scientific Approach.* McGraw-Hill Insurance Series, New York, 1931.
28. Taxis, K., Gallivan, S., Barber, N. and Franklyn, B. *Can the Heinrich Ratio Be Used to Predict Harm from Medication Errors?* Report to the Patient Safety Research Programme, U.K. Department of Health, University of Birmingham, West Midlands, U.K., 2006.
29. U.K. Parliament. Health and social care act 2012. http://www.legislation.gov.uk/ukpga/2012/7/enacted (accessed August 21, 2013).
30. European Commission. The Medical device directives: 2007/47/EC and 93/42/EEC (medical devices), 90/385/EEC (active implantable medical devices), 98/79/EC (in vitro diagnostic devices). http://ec.europa.eu/health/medical-devices/documents/index_en.htm (accessed August 21, 2013).
31. U.K. Parliament. Consumer protection act 1987. http://www.legislation.gov.uk/ukpga/1987/43 (accessed August 21, 2013).
32. Barker, R. and Storey, C. *Health & Safety at Work*. Tolley Publishing, London, U.K., 1992.
33. Health and Safety Executive. INDG291 (rev1). Providing and using work equipment safely – A brief guide. March, 2013. http://www.hse.gov.uk/pubns/indg291.pdf (accessed August 21, 2013).
34. U.K. Statutory Instruments 1998 No. 2306. Provision and use of work equipment regulations 1998 (PUWER). http://www.legislation.gov.uk/uksi/1998/2306/made (accessed August 21, 2013).
35. U.K. Statutory Instruments 1992 No. 2793. The manual handling operations regulations 1992. http://www.legislation.gov.uk/uksi/1992/2793/contents/made (accessed August 21, 2013).
36. U.K. Statutory Instruments 1998 No. 2307. The lifting operations and lifting equipment regulations 1998. http://www.legislation.gov.uk/uksi/1998/2307/contents/made (accessed August 21, 2013).
37. U.K. Statutory Instruments 1989 No. 635. The electricity at work regulations 1989. http://www.legislation.gov.uk/uksi/1989/635/contents/made (accessed August 21, 2013).
38. MHRA. Medical electrical installation guidance notes (MEIGaN). 2007. http://www.mhra.gov.uk/home/groups/comms-ic/documents/websiteresources/con2018069.pdf (accessed August 26, 2013). Under review.
39. European Commission. Waste Electrical and Electronic Equipment (WEEE) Directive 2012/19/EU. *Official J. Eur. Union*, L 197/55, pp. 38–71, 24 July 2012.
40. U.K. Parliament. Health act 2006. http://www.legislation.gov.uk/ukpga/2006/28/contents (accessed August 21, 2013).
41. ICRP. The 2007 Recommendations of the International Commission on Radiological Protection. ICRP Publication 103. *Ann. ICRP*, 37(2–4), 1–332, 2007.
42. U.K. Health Protection Agency. Application of the 2007 recommendations of the ICRP to the UK. 2009. http://www.hpa.org.uk/webc/HPAwebFile/HPAweb_C/1246519364845 (accessed August 21, 2013).
43. European Council. 1996. Directive 96/29/EURATOM. Basic safety standards (BSS) directive. http://ec.europa.eu/energy/nuclear/radioprotection/doc/legislation/9629_en.pdf (accessed August 21, 2013).

44. U.K. Statutory Instruments 1999 No. 3232. The ionising radiations regulations 1999. http://www.legislation.gov.uk/uksi/1999/3232/made (accessed August 21, 2013).
45. U.K. Parliament. The radioactive substances act 1993. http://www.legislation.gov.uk/ukpga/1993/12/pdfs/ukpga_19930012_en.pdf (accessed August 21, 2013).
46. U.K. Statutory Instruments 2010 No. 675. The environmental permitting (England and Wales) regulations 2010. http://www.legislation.gov.uk/uksi/2010/675/made (accessed August 21, 2013).
47. U.K. Statutory Instruments 2000 No. 1059. The ionising radiation (medical exposure) regulations 2000 (IRMER). http://www.legislation.gov.uk/uksi/2000/1059/contents/made (accessed August 21, 2013).
48. U.K. Statutory Instruments 2005 No. 2686. The high-activity sealed radioactive sources and orphan sources regulations 2005. http://www.legislation.gov.uk/uksi/2005/2686/contents/made (accessed August 21, 2013).
49. U.K. Statutory Instruments 2002 No. 2677. The control of substances hazardous to health regulations 2002 (COSHH). http://www.legislation.gov.uk/uksi/2002/2677/made (accessed August 21, 2013).
50. Health and Safety Executive. 2013. Control of substances hazardous to health (COSHH). http://www.hse.gov.uk/coshh/ (accessed August 21, 2013).
51. U.K. Statutory Instruments 2009 No. 1348. The carriage of dangerous goods and use of transportable pressure equipment regulations 2009. http://www.legislation.gov.uk/uksi/2009/1348/contents/made (accessed August 21, 2013).
52. U.K. Statutory Instruments 1995 No. 3163. The reporting of injuries, diseases and dangerous occurrences regulations 1995. http://www.legislation.gov.uk/uksi/1995/3163/made (accessed August 21, 2013).
53. Health and Safety Executive. Bed rail risk management. Sector Information Minute (SIM 07/2012/06). http://www.hse.gov.uk/foi/internalops/sims/pub_serv/07-12-06/ (accessed August 21, 2013).

# 4

## Approaches to Equipment Management: Structures and Systems

### 4.1 Introduction

Managing medical equipment in a large healthcare organisation is a major undertaking. It requires dedicated planning and organisation to be effective. In this chapter, we look at the management structures, administrative arrangements and operational capabilities an organisation needs before considering who might provide the major functions of financing, equipping, operating, maintaining, supporting and replacing equipment.

The focus of medical equipment management is naturally on the equipment itself – the operational tasks of getting new equipment into service, looking after it and finally replacing it as old equipment is withdrawn, as set out in the equipment life cycle in Chapter 2. However, various other management, administrative and governance processes are essential to make sure medical equipment is used safely and effectively. Each healthcare organisation has its own unique way of organising and carrying out these processes but every element is needed to support equipment effectively through its life cycle. In addition, the organisation must choose how much equipment management activity it will undertake directly and how much to buy in. At an operational level, almost any aspect of equipment management can be provided either in-house or by an external provider. At one extreme, some organisations outsource all equipment management to an external organisation and at the other, everything is done by their own staff. In practice, most organisations with a significant amount of equipment operate a mixed economy.

Even where operational equipment management is fully outsourced, final responsibility for governance and contract management still rests with the healthcare provider (see Chapter 8). It must retain strategic control of the equipment management process, overall risk management and service quality monitoring. It may also need to employ or commission experts to monitor its external contracts and risk management arrangements. An organisation can protect itself to some extent by careful wording of contracts and suitable insurance but cannot outsource overall responsibility for issues such as liability for adverse incidents, failure to meet operational objectives or inefficient use of resources.

In contrast, practical experience suggests that many day-to-day operational support and maintenance functions can be carried out more cost-effectively using in-house expertise, particularly in large organisations. However, a total reliance on in-house support for all equipment would be astronomically expensive in training and stocks of spare parts, and in-house staff would lack the depth of knowledge and experience available to suppliers. In-house services clearly depend on manufacturers, even if only for the supply of information, equipment and spare parts. Thus, in practice, equipment management is a dynamic partnership between an organisation and its external suppliers. The decision between whether to support equipment in-house or to outsource operations is a complex and dynamic strategic and operational balance that changes as equipment and staff come and go and as national regulatory and fiscal policies and practices change. Each organisation will have its own historical background of service planning and provision leading to a particular mix of in-house and external provision that, in the absence of major external forces for change, can evolve organically and adopt self-driven improvements. It is appropriate to test an organisation's business model from time to time, to consider if cost-effectiveness and quality could be improved through further changes to the balance of service provision, as discussed in Chapter 9.

## 4.2 Organisational Structures to Support Medical Equipment Management

### 4.2.1 Introduction

In this section, we describe and consider the strategic and administrative structures within organisations that support and deliver their equipment management processes. The schematic organisational chart in Figure 4.1 and supporting Table 4.1 identify the essential medical equipment management functions carried out at different levels in a healthcare delivery organisation and illustrate a typical equipment management structure. From the top of the chart downwards, the bias of committee and group membership moves from non-executive directors towards senior operational personnel. The division of responsibilities, the name and membership of each group and indeed whether or not they exist at all will vary depending on the size and nature of the organisation. All the responsibilities adopted by the groups as described here must be allocated however to a committee or person at some level in the organisation. The point of balance between freedom for clinical users and centralised control of risks and resources will depend on the culture and aims of the organisation and the skills available to it.

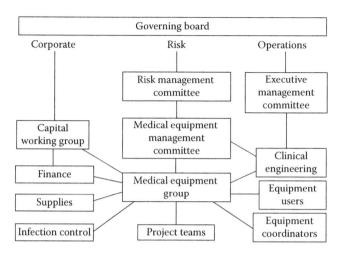

**FIGURE 4.1**
Organisational structures to support medical equipment management.

## 4.2.2 Governing Board

A healthcare organisation of any size will have a board of directors or trustees responsible for governing it, a task that involves deciding on long-term aims, objectives and strategies and setting up the organisation to deliver the desired outcomes. Membership of this body will include executive directors, with specific managerial functions, and non-executive directors with particular skills who provide a level of external knowledge, scrutiny and challenge over strategic direction, operational decisions and overall organisational performance. Executive, management and operational functions are devolved from the board to various individuals, committees, groups and services.

Medical equipment will come to the attention of the board in three principal ways. First, it will appear as strategic enabler, supporting the provision and development of new and existing clinical services. Investments in large equipment projects often have a high organisational and public profile, affecting the provision of core clinical activities, setting the direction of service developments in anticipation of future patient demands and contributing to the public image of a successful organisation. The board will be instrumental in setting and approving strategic direction and capital investment but will devolve specific projects involving medical equipment to an operational or project group. It is also likely to require periodic progress reports for larger projects and to be notified generally of any matters of significant concern. Secondly, the supply and care of medical equipment is likely to be a major source of expenditure. Raising money to purchase equipment, and budgeting for its maintenance and support, will be a major concern for clinical users and the organisation's finance function and the board will want to know that these costs are correctly identified and monitored and can be met.

**TABLE 4.1**

Groups Relevant to Equipment Management in a Typical Organisation

| Group | Key Areas Represented | Medical Equipment Functions Include | Role of Clinical Engineering |
|---|---|---|---|
| Board | Executive and non-executive directors | Major project decisions, capital allocations | Papers and briefings |
| Operational management executive committee | Executive and clinical directors, finance, HR | Approve smaller projects, policies and strategies | Write and present cases |
| Capital programme committee | Directors responsible for finance, IT, estates, etc. | Coordinate capital funding, recommend capital allocations | Report on replacement/ purchasing |
| Medical equipping group | Medical director, clinical operations, finance, estates, IT, clinical engineering | Prioritise bids for funding, oversee purchasing | Provide detailed support |
| Risk management group | Medical director, senior clinical staff and other experts | Oversee governance, risks, safety and policies | Reports, investigations |
| Medical equipment management committee (MEMC) | Doctors, pharmacy, nurses, IT, estates, finance, procurement, clinical engineering | Identify, investigate and propose actions to improve safety and reduce risk | Detailed support to committee |
| Other groups such as equipment managers and device trainers | Clinical engineering and others as appropriate | Improve operational management, reduce risk | Facilitation, advice |
| Project teams/ working groups | As appropriate | Delivery of particular projects | Specialist input |
| Other relevant committees | Clinical engineering specialists | Expert advice, coordination | Provide advice |

Most organisations have an internal audit function, reporting directly to the board, that will take an in-depth look at value for money in medical device activities such as procurement, maintenance and disposal. Finally, equipment will be seen as a source of risk, whether financial, operational or legal. This aspect is usually monitored through a dedicated committee that establishes an overall risk management framework, to devolve individual risks to risk management groups in the organisation.

The committees and groups directly involved with medical equipment management may report to the board in various ways but all aspects of medical equipment care will in practice need to come together at some level below the board itself. Key individuals and groups using medical devices may belong to one or several of these groups and may report to others on a regular or ad hoc basis. A single executive director should take responsibility for medical device issues across the organisation, acting as overall sponsor for medical device management and providing strategic leadership. This director can sponsor business cases at the board and in other management groups and act as a champion for improving patient safety and equipment management procedures and processes. Ideally, this role is taken by the senior managing clinician (medical director), who can balance clinical priorities against resource issues. It is a role sometimes taken by the director of estates and facilities or head of clinical engineering, whose stronger emphasis on resource management will need to be supported by appropriate clinical input.

### 4.2.3 Operational Executive Management Committee

This is the committee to which the board delegates responsibility for operational management. It is pivotal to the successful running of the organisation and is the forum where executive and clinical directors set detailed strategy and policy, monitor performance and make key decisions on business and clinical strategy across the organisation in the light of the overall objectives and strategies set by the board, service needs, internal and external changes and financial and other pressures. It receives information from all groups within the organisation and has the best integrated overview of healthcare delivery. This committee may be supported by a group drawn more widely from senior clinicians, senior managers and heads of some specialist services. For equipment management issues, this group will advise on major operational matters including investment in large or expensive equipment programmes and is likely to be responsible for ratifying the organisation's medical device policy, with suitable input from the risk management committee and other appropriate groups. It is therefore the point at which tensions between clinical risk management and resource issues come together to be resolved. The two strands of resource management and risk management tend to follow separate management lines from this group downwards.

### 4.2.4 Risk Management Committee

This group and its supporting functions coordinate the assessment of risks to patients, staff and the organisation and monitor how well these risks are managed by the operational side of the organisation. Part of its remit will be to investigate accidents and incidents arising within the organisation, to make sure lessons are learned and problems are addressed, analysed and reported. It maintains a register of risks on behalf of the organisation and ensures that priorities are set and resources identified to address them. It will make sure that legislative and advisory requirements are implemented and will report to the board on progress in managing and reducing risk. Its remit will cover health and safety risks of all kinds, with the most common categories being issues such as slips, trips and falls (by both staff and patients) and medication errors, in addition to problems arising from medical devices. The emphasis in this committee is more on patient safety and the monitoring and prevention of untoward incidents than with the financial and operational efficiencies of the organisation. One important function of this committee regarding medical equipment will be the identification and promotion of urgent equipment procurement or replacement needs in response to untoward incidents or identified risks.

### 4.2.5 Capital Programme Committee

The organisation's finance director will report to the operational management committee and the board regarding the organisation's financial position and future capital and revenue projections. Various groups will provide information and feedback to support the allocation of monies and help to keep track of income and expenditure.

Capital funding is often coordinated separately to revenue, with a capital programme group overseeing capital allocation and expenditure across a range of programmes such as estates, information technology (IT) and medical equipment (Chapter 5). Membership of this group is likely to involve lead directors from each of these areas, along with finance and other specialists who can provide the specific expertise needed to monitor and direct the capital programme. It is worth remembering that the sums of money involved in medical equipment purchases, which may seem large to an individual clinical engineer, usually represent less than 10% of the total assets of the organisation when buildings and major plant are taken into account. Nevertheless, the financial and governance processes associated with equipment will be audited and monitored as closely as any other aspect of expenditure. Each capital budget – estates, IT, medical equipment – should be allocated to various projects by a group able to judge between and prioritise different calls for funding.

### 4.2.6 Medical Equipment Management Committee

This committee may report directly to the board or go through the risk management committee or other bodies such as the operational executive committee. Chaired by a senior clinician, lead director or clinical engineer, the focus of its activity varies from resource allocation to devising and recommending medical device policy. It sits at a level in the organisation where its members will be involved in the detailed scrutiny of both risks and resources. Some medical equipment management committee (MEMC) groups make major strategic equipment management decisions such as prioritising the annual programme of capital medical equipment projects. It consists of people at a sufficient level of authority to be able to decide on the relative merits of different resource investments, in the light of clinical need, and who can implement decisions on policy and risk management. Membership typically might include lead clinicians, directorate managers, representative equipment managers and other lead clinical engineers and medical personnel including infection control staff. Senior staff from clinical engineering usually supports its activities.

### 4.2.7 Supporting Groups

Other groups that might support this structure include the following.

#### 4.2.7.1 New Devices Group

New devices are regularly introduced into healthcare organisations, often through a variety of unrelated routes and in tandem with new clinical procedures. Keeping control of the risks and costs this presents is a challenge to any dynamic organisation. Rather than leaving the risk issues to a generic risk management committee and financial concerns to the individual services, some organisations may set up a specific group to approve new or changed device types, consumables or clinical practice. The group is likely to comprise nominated clinical, supplies and clinical engineering staff. The aim of this group will be to ensure that proper consideration has been taken of clinical risks, liabilities, standardisation and ongoing costs when new items are introduced, whilst ensuring that innovation is effectively managed and implemented. They may undertake cost–benefit analysis of new or updated technology.

#### 4.2.7.2 Project Teams

These are set up for any non-trivial change, with membership drawn from all affected groups. Whilst the scope of projects might vary, from purchase of an individual item to equipping a whole hospital department, similar project management techniques can be applied. Project teams can be set up by, and report to, any of the groups identified earlier. Organisations may require that all major projects report to the operational management group. Financial criteria might be applied for deciding whether or not a formal business plan is required, and as a general rule, no project significant enough to require a formal business plan should be implemented without investment in a supporting project team. The project team will report to a project sponsor or steering group and will be run by an identified project manager overseeing a team with appropriate expertise. The team will typically produce the business case, arrange the tender, oversee device evaluation and procurement and ensure appropriate arrangements are made for delivery, commissioning and associated issues such as building works.

#### 4.2.7.3 Clinical Engineering

Clinical engineering is the service that provides the majority of day-to-day technical management of equipment, together with expert advice on strategic equipment management and policy. Its contribution can include advice on procurement and drawing up specifications; assistance with evaluations and clinical trials; acceptance testing; user training; management of the equipment inventory; routine and breakdown maintenance; condemning and disposal of unwanted equipment; investigation of incidents and the provision of technical and scientific advice; clinical support for research; and service contract management. Senior clinical engineers may have strategic and executive roles in the organisational management structure and may report directly to interested committees and groups, particularly the MEMC, whilst also being incorporated in the management structure in areas such as clinical support or estates and facilities. The detailed structure and operations of clinical engineering services is described in Appendix A.

### 4.2.7.4 Users

Equipment users vary considerably in their knowledge of, and engagement with, medical equipment management. One way to organise medical equipment support across a large organisation is to set up a network of equipment coordinators, who are end users with particular responsibility for the local management of equipment. These responsibilities might cover equipment in a specific location, of a specific type or used by a particular clinical service. Also various specialist clinical scientific and technical groups are typically responsible for operating complex equipment and are likely to have developed expertise in looking after their own front-line maintenance, quality assurance, patient training and service contract management. They are typically involved in operating imaging or therapy devices using ionising or non-ionising radiation, in physiological measurement or monitoring, in rehabilitation engineering, in critical care or in theatres. The clinical engineering department can assess equipment management needs in different areas and help equipment coordinators achieve value for money in equipment management across the organisation.

### 4.2.7.5 Organisation-Wide Lead Roles

Many smaller organisations have a single designated equipment manager who functions as the organisational lead for strategic aspects of medical equipment management. The person or persons performing this role will be responsible for the following:

- Leading on organisation-wide risk and governance issues for medical equipment
- Writing the medical device and equipment management policy for the organisation and developing strategies for its implementation
- Facilitating compliance across the organisation, through suitable advice and monitoring

- Coordinating the organisation's overall medical equipment procurement programme
- Sitting on related groups, such as decontamination or point of care testing
- Providing the visible face of equipment management to the organisation and externally
- Overseeing external equipment support contracts

It is important to recognise the need for leadership in these roles, otherwise the organisation will lose focus. Effective support from a lead director at board level is crucial in maintaining an organisation's attention on medical device issues and medical equipment management.

## 4.3 Systems for Equipment Management: Balancing In-House and External Provision

Having reviewed the organisational structures needed to control the management of medical devices, we now consider options available for organising delivery of various equipment management functions. Whatever model is chosen to provide equipment financing and management, the organisation has to maintain overall control using some variant of the structures described earlier, scaled to suit the size of the organisation and tailored to operate effectively with externally supplied services. We begin consideration of this topic by looking at what happens in major new hospital projects.

### 4.3.1 Financing and Equipping Major Projects: An Overview

Large-scale changes to a health system, such as the building of a new hospital, amalgamation of existing hospitals, addition of a new hospital building or changes to a major clinical service on an existing site are likely to involve large building and equipping costs. How these costs are funded has long-term consequences for a healthcare organisation and constrains its freedom to organise equipment management services. For example, if capital for financing and equipping a new development comes from private sources through a public–private–partnership (PPP) model (see Section 4.3.2) or an equivalent scheme, ongoing equipment maintenance and replacement programmes are likely to be paid for under the same arrangement. If money from the organisation itself, government funding or charitable sources is used to finance the project, there is more flexibility but less certainty about future arrangements for equipment maintenance and replacement.

To explore the issues involved in these choices, we now consider the two distinct models of PPP and in-house equipping as applied to large projects.

Operational equipment management in a major project can be summarised under three main headings: financing and equipping, operation and maintenance and ongoing equipment replacement. For any one of these areas, there are multiple ways to set up a balance between predominantly in-house provision and provision by an external private or other public or non-profit provider. Making choices in one area may however restrict or influence future decisions in another. For example, setting up an equipment replacement programme funded by regular payments usually involves some restriction on the choice of makes and models of equipment that are provided, if the contract is to be competitive economically. Some of the complexity of this area is explored in the following text.

### 4.3.2 Public–Private Partnerships and Equivalent Schemes

Public–private partnership or PPP schemes – also known as private finance initiatives (PFI) in some countries – have been used to provide infrastructure for public sector undertakings such as hospitals, schools, prisons and transport, initially in Australia and the United Kingdom and then spreading to Mainland Europe, Japan and an increasing number of other countries. Under these schemes [1–3], a consortium of financiers, builders and manufacturers set up a dedicated company as the PPP service provider to finance, build and equip the hospital using private sector capital. On completion, the organisation leases the buildings and equipment from the PPP partner for a long period, typically 30 years. During this time, the PPP partner manages and maintains the infrastructure and may also manage medical equipment in addition to managing or subcontracting facilities such as portering, catering, security and IT. The PPP buildings and any equipment funded through the scheme continue to be owned by the PPP consortium whilst they operate the facility and usually become the property of the organisation at the conclusion of the scheme. It is common to separate the facility from the equipment and to use an in-house service or a separate PPP partner for all or part of the equipping. One trend is for partial equipment provision where the PPP company provides large and expensive items of equipment but leaves the provision of smaller items to the organisation.

PPP schemes have been actively encouraged, often with partial or total exemption from elements of taxation on the costs of managed services and facilities [4]. An argument put in favour of PPP is that outsourcing of infrastructure and facilities allows an organisation to concentrate on its core business of providing healthcare. However, even given a taxation advantage, the total cost of a PPP deal usually exceeds that of the publically funded alternative [5]. Reasons for this include the need to provide a return on capital for investors and high costs levied by the PPP provider to make alterations to the building or equipment, in response to the not infrequent changes in healthcare delivery required in an active healthcare provider. The accounting device of discounting present net value aims to correct costs involved for the effects of inflation over the time of the scheme [6], and this too can lead to distortions as inflation is not uniform in the economy – for example, medical equipment costs have been increasing more slowly than general inflation, so that equipment replacement programmes are overall becoming cheaper in real terms for a similar relative level of technological sophistication. Schemes which build inflation into equipment values can therefore make substantial additional profits if the healthcare organisation using them does not set up or monitor contracts appropriately.

All models of outsourced service provision claim to transfer risk from the public to the private sector. This is true for the building phase of PPP projects, where the costs of project management failures – running over time or budget, or not providing the agreed facilities – are borne by the PPP provider. During a building's life, however, risk transfer is less clear, with the private partner guaranteeing a level of service rather than underwriting the clinical consequences of any service failure. Contracts usually include financial compensation to the healthcare organisation for sub-standard services but the organisation is still the front line for any penalties arising from any consequences to patient care or a failure to meet healthcare delivery targets and cannot divest itself of all risks [7]. Where projects have totally failed [6] and the associated company has been wound up, the costs and responsibilities involved have reverted to national governments.

The complexities of PPP negotiations and service specifications require specialist knowledge. A healthcare organisation will want to appoint consultant equipment advisors to work alongside the in-house clinical engineer, to produce a general policy and a detailed output specification for the project. Where large projects are outsourced, specifications for every aspect of the project must be drawn up in considerable detail and the cost and manner of provision negotiated carefully. If existing equipment is to be included, an accurate and detailed inventory will be required with data on present value and expected lifetime. This information acts as a baseline for specifying the quantity and type of equipment required for the project and identifies equipment with sufficient remaining operational life and value to be worth transferring to the new development. The inventory will also be used as a basis for developing an equipment replacement programme, where this is to be controlled by the private sector partner. Where new services are involved, functional and service level specifications for any additional equipment will need to be drawn up prior to procurement, making sure that services in the new project are funded to support it. Advice from professional colleagues with experience of such schemes is invaluable in helping to prepare for and respond to the challenges of such a project. This particularly applies when existing in-house services, including technical staff, may be transferred to a PPP facilities management service under the transfer of public undertakings (TUPE) [8] regulations. The healthcare organisation will need to retain

**TABLE 4.2**

Equipment Groups

| Equipment Groups | Examples of Equipment |
|---|---|
| 1. Fixed infrastructure | Lifts, air conditioning, IT networks, power distribution |
| 2. Large, high value | Radiology equipment, laboratory analysers |
| 3. Medium value, portable | Ventilators, patient monitors, near patient testing |
| 4. Small cost and size | Oxygen regulators, nebulisers |

knowledgeable individuals or otherwise obtain expertise to assist in contract monitoring and other areas of governance, both clinical and financial. Areas needing particular attention include equipment-related incidents, risk management, contract performance and value for money on equipment purchases and contract variations.

For the purposes of major equipping projects, particularly those involving PPP, it is helpful to classify equipment into groups that reflect the size, portability and purchase cost of the various items. This then directs how equipment procurement and management processes are best applied. These groups are summarised in Table 4.2. From a day-to-day administrative point of view, the cost and time spent managing and tracing small items is disproportionate to their purchase cost. Large fleets of small items, for example, syringe drivers and medical gas flow meters, are subject to constant movement and handling and may be used by many different people. They can get lost in drawers and cupboards, be broken and put out of sight or leave a hospital altogether attached to patients. Battery-backed syringe drivers require charging and items need decontaminating and inspecting before reuse. In contrast, the effort involved in managing a CT scanner support contract, if all goes well, is relatively straightforward and it is unlikely to be lost! External service providers are therefore wary of taking on low-value equipment and may seek to exclude equipment below a stated purchase price from being included in a managed service or PPP deal, or charge a significant sum for managing it. So while it is almost certain that the equipment infrastructure items in group 1 will be supplied and managed by the PPP partner, group 4 equipment will usually be funded, procured and supported by the healthcare organisation, often at ward or departmental level. Between these extremes, there is wider scope for different approaches to devices in groups 2 and 3. The organisation must take strategic decisions as to whether to include this equipment in the PPP contract, commission a separate third party provider or take on responsibility for its funding, procurement and management. The high cost of group 2 equipment can make a variety of funding approaches attractive instead of outright purchase, including leasing and the use of managed services to spread purchase costs. For this equipment, maintenance and support is most likely to be provided by the original manufacturer or a third party provider. An organisation is more likely to be able to afford to buy group 3 equipment, and may also provide a substantial proportion of its maintenance and support.

### 4.3.3 Financing and Equipping by the Healthcare Organisation

In this model, the healthcare organisation is responsible for finding funding to build and equip the facility, from public or other sources. This can range from a complete public sector project funded from direct government sources or other dedicated public capital to a reliance on charitable grants or private sector loans. Large schemes may include a mixture of funding methods, including drawing on sources such as the sale of unwanted buildings. Even where a new build is government financed, the healthcare organisation is likely to provide some contribution towards equipping costs and will take over responsibility for long-term maintenance and replacement. Equipment procurement and maintenance funding may come from existing operational budgets. The organisation may also lease equipment and maintain it under the provisions of the lease or through in-house or third-party agents. Other schemes include rental or long-term purchase of equipment paid for on a cost per use or cost per test basis, or by an additional charge on consumable costs, and may include additional equipment management features aimed at reducing the direct cost and effort of equipment ownership. These schemes can reduce costs for an organisation but only if they are monitored and managed carefully and have contractual arrangements that provide flexibility to adapt to changing healthcare needs.

Operational equipment management is also likely to include service, quality or research functions not easily specified or readily delivered under a PPP contract. These include ongoing clinical governance support, the provision of research and development advice to clinical staff, associated in-house equipment modification and development facilities and innovations in service itself. Retaining maintenance service and support functions within the organisation avoids having to identify and cost out these activities separately, which suits an environment where demand swings between providing routine service and working on clinical development projects. Under these circumstances, the cost of in-house technical experts can be spread over a range of activities to deliver outcomes at lower rates than external consultancies hired to carry out a specific project. The flexibility of in-house support, when effectively used, can significantly cut servicing costs compared to external contracts but this requires effective management and control (see Chapter 9).

### 4.3.4 Managed Equipment Services

Where a healthcare organisation provides funding for equipment replacement, it may decide to outsource the management of ongoing capital and revenue replacement programmes to an external provider together with the provision of maintenance and management of contracts [9]. In a similar way to a PPP-managed equipment service (MES), an independent MES provider will be expected to provide an agreed number of items of equipment working to their functional specification, within uptime and call out response time targets. The scope of MES services ranges from whole hospital equipping to the provision of specific items of specialist and expensive equipment, with schemes to fund and maintain imaging and radiotherapy equipment being amongst the most common. They are attractive to healthcare organisations which are short of capital and where the installed facility is either essential to clinical operations or will generate a good return for the healthcare provider. The MES provider must have access to a source of capital and be capable of providing or commissioning maintenance and maintenance contract support.

The most straightforward concept for the management of maintenance alone is that of an external service provider, where a single company acts as a multi-vendor service (MVS) to manage service contracts and breakdown repair. Like PPP schemes, the contract may transfer existing staff from the healthcare organisation to the MVS. It is also becoming more common for manufacturers to take on equipment management and maintenance responsibilities which include other manufacturers' devices. This can involve a complete service, cover equipment of a certain type such as ultrasound scanners or operate across a defined service area such as intensive care. The advantage to a healthcare organisation, as with most outsourcing, is that it is perceived as a one-stop solution which allows it to concentrate on its core business. Contracts require careful negotiation, however, since much that is implicit in an in-house service will not be provided unless explicitly stated. Overall responsibility for clinical risk remains with the healthcare organisation even though the MVS takes over most service risks. Risk transfer has to be substantial for certain tax exemptions to apply [10].

Management of maintenance alone is more common in small organisations or group practices where the amount and value of equipment is relatively small. This type of service may be provided by a private sector company or not for profit organisation including the in-house service of a public sector healthcare organisation. A larger organisation may contract for managed services either piecemeal, involving different providers in specific areas where this has been found to be cost-effective, or may opt for an MES either as a stand-alone package or alongside an infrastructure PPP scheme provided by a different company.

In the United Kingdom, although there are some outsourced maintenance operations in existence that do not form part of a PPP scheme, most large healthcare organisations manage equipment flexibly via an in-house clinical engineering service. This service takes responsibility for the oversight of equipment management but relies on a mixture of manufacturers' service contracts, third-party contracts, MVS provision and in-house maintenance and repair. A minimal top-down department may manage and oversee replacement programmes, maintenance contracts and risk management (safety alerts and recalls, incident investigations and so on). Some clinical engineering services take considerably more responsibility for the implementation of routine maintenance, breakdown repair and front-line troubleshooting than others, with skilled staff employed in-house. Financial and operational factors relevant to a decision whether to perform these functions internally or outsource them are highlighted and discussed in Chapters 8 and 9 and Appendix A. Considerable knowledge and insight is required from technical experts to unearth the true costs and benefits of an in-house service or external provision. Where any operation is outsourced, controls and monitoring of the process are still required, and it cannot be stated emphatically enough that contract arrangements need detailed and regular scrutiny and active management if value for money is to be achieved and risk transfer of equipment failure is to be effective.

## 4.4 Summary

Medical equipment has multiple strategic, financial and risk management challenges and requires a number of organisational structures to meet them, including dedicated groups and effective communication. An organisation will need to have at least one medical equipment management group or function to oversee equipment planning and resource allocation. Ultimately, the healthcare organisation is responsible and adopts the risk for all its equipment management activities but will usually choose to outsource some functions and perform others in-house. We considered some ways in which this might be done, including Managed Equipment Services and Public-Private Partnerships. The exact mix of providers will vary considerably between organisations and must be decided, and periodically reviewed, by careful and detailed consideration of the implications for finance, clinical services and risk management.

## References

1. Sussex, J. Public–private partnerships in hospital development: Lessons from the UK's private finance initiative. *Res. Healthc. Financ. Manage.*, 8(1), 59–76, 2003.
2. Grout, P. The economics of the private finance initiative. *Oxf. Rev. Econ. Pol.*, 13(4), 53–66, 1997.
3. Grimsey, D. and Lewis, M. *Public Private Partnerships: The Worldwide Revolution in Infrastructure Provision and Project Finance*. Edward Elgar Publishing, Northampton, MA, 2007.
4. U.K. House of Commons. 2002–2003. Select committee on public accounts – twenty-eighth report. http://www.publications.parliament.uk/pa/cm200203/cmselect/cmpubacc/764/76404.htm (accessed on August 19, 2013).

5. Pollock, A., Shaou, J., and Vickers, N. Private finance and "value for money" in NHS hospitals: A policy in search of a rationale? *BMJ*, 324, 1205–1209, 2002.

6. U.K. House of Commons Economic Affairs Committee. 2009. Private finance projects and off-balance sheet debt. Letter and memorandum by the British Medical Association (BMA). http://www.publications.parliament.uk/pa/ld200910/ldselect/ldeconaf/63/09121507.htm (accessed on August 19, 2013).

7. Burge, D., Bingham, C., and Lewis, A. 2012. Risk transfer in outsourcing contracts. http://crossborder.practicallaw.com/2-518-7949 (accessed on August 19, 2013).

8. Department for Business Innovation and Skills. 2012. 09/1013 – Employment rights on the transfer of an undertaking, A guide to the 2006 TUPE regulations for employers, employees and representatives. https://www.gov.uk/government/uploads/system/uploads/attachment_data/file/14973/2006-tupe-regulations-guide.pdf (accessed on September 06, 2013).

9. Lansdown, S. and Thomas, S. Public-Private Partnerships: Getting NHS Finance That Adds Up. *Health Service Journal*, 23rd October 2009. http://www.hsj.co.uk/ (accessed on September 06, 2013).

10. HM Treasury. A new approach to public private partnerships. Crown copyright, 2012. https://www.gov.uk/government/uploads/system/uploads/attachment_data/file/205112/pf2_infrastructure_new_approach_to_public_private_parnerships_051212.pdf (accessed on August 19, 2013).

# 5

## Purchase and Replacement: Allocating Priorities and Managing Resources

### 5.1 Introduction

Clinicians are rarely content to carry on using the same medical equipment year after year. Even where services are relatively well equipped, this year's technology usually offers more than last's. Clinical aspirations and patient expectations, driven by pressure to replace or upgrade existing equipment to improve patient treatment, safety and service efficiency, create a demand for equipment funding that is hard to satisfy. There are always limits on the amount of money an organisation can invest in its medical equipment, yet limits are not always a bad thing. They direct an organisation to look at what it really needs and encourage it to get the most out of what it already has. Many healthcare organisations are not good at doing this: for example, a surprising number of clinical and capital developments are initiated without considering the full consequences on future equipment needs and spending [1].

This chapter addresses the question: 'How do you allocate finite resources to purchasing medical equipment?' We consider factors driving the purchase of medical equipment, and then describe processes by which healthcare organisations can identify their equipment replacement needs and match these to available resources. We look at ideas for prioritising resource allocation, including a suggested model for capital allocation decision making that provides organisational oversight of local priorities. Having accepted the need for a purchase to go ahead, there is then a further decision on which available funding stream to use – capital, revenue, charitable – and the procurement model to follow, such as purchase, lease, hire or managed service. We outline how the procurement process is implemented in order to provide effective use of resources and ensure that equipment, once purchased, is fit for purpose.

This chapter is not concerned with lower cost equipment purchases which are built into local budgets. It is relevant when competing for prioritisation in any local, organisational or national medical equipment resource allocation. Organisations which never set conditions on or say "no" to an equipment purchase proposal are either fortunate or profligate, depending on one's point of view. They are also unlikely to exist for long.

---

## 5.2 Seeking the Ideal: Matching Needs and Resources

In order to devise and manage equipment replacement and development programmes efficiently, an organisation must be aware of the current and future funding available to it, including different funding methods and any flexibility to convert between one funding stream and another. It must also be aware of its current and future needs for equipment, have a way to review and change medium- and long-term priorities in reaction to changing clinical requirements and technologies, and be able to react swiftly to emergencies. In short, it requires an effectively managed and comprehensive process that links board decision making to clinical and strategic priorities. Even where this ideal is achieved, external circumstances can disrupt even the best-planned programmes. Funding levels vary unpredictably from year to year, health policy and regulations can introduce whole new areas of equipment requirements, and those responsible for clinical equipment may be more concerned with using it than planning for its cost-effective replacement. Under these circumstances, equipment replacement becomes a complex and interactive political process.

Managing clinical expectations and the politics of resource allocation is demanding. Organisations steer a line between central and local decision making, whilst being buffeted by urgent clinical demands. Less central effort means more reliance on effective local cooperation between managers and clinicians. The difficulty of making value-based decisions between very different types of equipment across the organisation, in the light of multiple competing clinical needs, is a worthy challenge and requires good central oversight and a regular re-examination of the allocation processes if it is to be done effectively. Ideally, organisations should look critically at full lifetime costs (capital, depreciation and running costs) when making investment decisions. Identifying these costs requires significant effort, particularly where technology is changing and developing more rapidly. The effective purchasing lifetime of many types of equipment is now too short to carry out a thorough clinical, technical and financial evaluation based on historical performance. Detailed analysis across a large number of items is time-consuming and difficult for a single organisation to do. These problems are compounded by the difficulty of predicting what functions clinical users need, rather than what they want, as user perceptions of what is vital can be significantly at odds with the way equipment is used in practice. The importance of a good funding allocation process for medical equipment becomes apparent when there is competition for resources with other major users of capital.

## 5.3 Funding Routes for More Expensive Equipment

In most hospitals, funding for medical equipment is determined in competition with demands for investment in IT systems, buildings and infrastructure. Although ultimately there are limits on the amount of capital and revenue funding available to any organisation, flexible and intelligent use of resources can maximise the amount of equipment that can be procured. How far down the priority list for funding the money will stretch depends partly on a creative mix of the various types of funding on offer.

### 5.3.1 Capital Funding – Definition

Although distinctions between what constitutes capital and revenue funding may vary from country to country or be non-existent, the United Kingdom is given as an example to form a basis for understanding funding principles and processes. In the UK, capital equipment is a physical asset with cost above a certain threshold, historically set at £5000 for the UK public sector, which also has a minimum expected lifetime, often taken as more than a year.

Capital is made available to public sector healthcare organisations through direct or indirect public funding, for example, by awarding annual grants, via a bidding process or for specific innovations or projects. Earned capital is generated in healthcare institutions not funded directly by government, such as independent hospitals or UK foundation trusts, where a proportion of operating profit needs to be allocated to invest in capital equipment.

Another opportunity open to independent organisations such as UK foundation trusts is to borrow capital funds from private or government sources against a specific development. Such projects need to have a robust business case, with the resulting income or other benefit being sufficient both to repay the loan and make a reasonable financial return.

### 5.3.2 Charities and Research Funding

Charitable grants invariably come with conditions attached. They are particularly appropriate for new equipment where the running costs of service developments are affordable but the capital costs are not. Many large items, such as CT scanners, have been purchased in this way. Sometimes, physical equipment, rather than the funding for its acquisition, is provided directly from charitable sources but this can be particularly difficult to manage if the items are not what the organisation itself would have chosen. This is a problem often found in developing countries.

Funding may be specifically allocated to purchase medical equipment for teaching or research, as part of an overall infrastructure or individual research project, through government, research funders, higher education or philanthropy. The organisation may be able to use this equipment to enhance healthcare delivery.

### 5.3.3 Revenue Funding

The range of revenue funding methods for capital items varies widely but usually involves some kind of hire purchase or equivalent scheme, where the supplier owns the equipment until sufficient payment has been made to cover its cost.

#### 5.3.3.1 Managed Service

One option is managed service, where a fee is paid per procedure and equipment remains owned by the supplier, who also manages its servicing and renewal. These services may be exempt from VAT or other taxation, which can provide a financial advantage under carefully controlled circumstances. However, it should not be assumed that tax relief will always be applicable, and it is important to explore the rules in detail as it will depend on the exact nature of the service provided.

Acquisition of equipment as part of managed equipment services or private finance initiative deals is discussed in Chapter 4. It is usually a major decision associated with fundamental restructuring of facilities and clinical services. These arrangements are set up for the long term with payments from revenue. However, the same principles regarding equipment replacement and renewal apply whatever the source and timescale of funding, including the need to explore alternative options and scrutinise contract terms carefully, to avoid being tied in to premature equipment replacement.

#### 5.3.3.2 Consumables Related

With equipment, such as large-scale laboratory analysers, the annual revenue budget for consumables, such as reagents or test kits, may be many times the capital cost of the equipment, and sometimes, the manufacturer will provide equipment for no initial cost if the organisation agrees to buy an annual minimum level of consumables. Maintenance and final ownership may also be included in such deals.

#### 5.3.3.3 Leasing

Capital equipment can be funded from revenue budgets rather than a capital pot by leasing, which spreads payment over time and converts a one off capital payment into an ongoing revenue stream. The equipment is purchased by a finance company to which the healthcare organisation then pays rental for use of the equipment for a period of typically 5–7 years. Healthcare organisations have a practical limit on how much capital they can lease, set by the market or by government, and need to generate sufficient revenue to support the investment. Leased projects require active management to ensure value for money, particularly to avoid excessive payments for short lease extensions at the end of the lease period. When capital is limited, the capital allocation process may also identify projects which are suitable for leasing, freeing capital for other items. Central control helps an organisation direct overall equipment funding, but if authority is delegated locally, lease purchasing becomes more difficult to control.

### 5.3.3.4 Loan and Hire

Hire or direct rental of equipment is often undertaken in response to equipment breakdown or sudden increases in clinical service demand. Other ways to obtain equipment for short periods include loans from manufacturers (often pending delivery of new or replacement equipment) or from other hospitals. There are also service organisations which specialise in providing a range of equipment for rent. Payment can either be by period of time or by number of patient uses. The ultimate form of this arises where a clinical service, usually in imaging or theatres, rapidly outgrows both its available equipment and space or when a local disaster renders facilities unusable for an extended period. A complete clinical service can then be brought in by private healthcare specialists, ranging from a mobile scanner to a fully equipped portable operating unit supported by clinical, administrative and scientific staff and rented until a long-term solution is put in place. Rental is an expensive option.

### 5.3.4 Using Capital Funds for Low Cost Items

Revenue items were defined above as costing less than a certain threshold – currently set at under £5000 in the United Kingdom – for each self-contained device. A certain amount will be built into local budgets for medical device purchase. Some items not classed as capital can be funded from capital money either by grouping together a number of interrelated items that cannot function without each other or by a capital to revenue funding transfer. The purchase of large numbers of independent items such as beds and couches would not fall into the first category but may come into the second, as would equipping elements of a whole new unit.

## 5.4 Identifying Equipment Needs

The main reasons for acquiring medical equipment are as follows: to replace existing equipment which has failed or is unreliable; to acquire new technology or additional equipment to develop services or reduce costs; to reduce risk, by measures such as standardisation or new technologies; to meet regulatory pressure and to provide specialised equipment for research.

### 5.4.1 Routine Replacement

Routine replacement keeps the equipment stock up to date. It is good practice to replace equipment before it becomes excessively unreliable or fails totally and disrupts clinical services. Gathering evidence on ongoing performance will improve planning and avoid wasting money on premature replacement. Replacement programmes require regular assessment of the residual life of equipment. The notional write off period adopted for accountancy purposes is often unrelated to usable operational life, as good design, lower intensities of usage, careful handling and good maintenance all tend to lengthen the effective life of individual items, whilst rough treatment by users, poor maintenance, defects in design and manufacture or technological obsolescence will shorten it. With limited resources, routine replacement may be a luxury and the routine replacement of equipment may be carried out only when significant breakdowns or failures occur or a manufacturer finally runs out of critical spare parts.

### 5.4.2 Replacement due to Unreliability

As equipment ages, elements start to fail, or fail more often. At first, repairs should be possible and be of an acceptable cost, though judgements must be made as to what repairs are economically viable. As time goes on, downtime is likely to increase and clinical services may be affected if insufficient back-up equipment is available. It is usually at this point that a planned equipment replacement programme begins to be implemented. There are two areas which are worth exploring before making the assumption that ageing is the primary cause of unreliability: operator misuse and consequent damage, and an unsuitable environment. For example, providing proper storage facilities and medical equipment training will often cut down the number of incidences of damage.

### 5.4.3 Failure

A point will eventually be reached where equipment stops working and repair is no longer economic. How old the equipment is when this happens often depends on the intensity and type of use and how well it has been looked after. However, as modern equipment becomes generally more reliable and component manufacturing lifetimes reduce, the manufacturer may cease to support equipment before this point is reached. It is advisable to consider what contingency arrangements will be put in place to cover equipment failure, as the service impact of these may show more clearly whether replacement is really needed.

### 5.4.4 Technological Development

New technology provides equipment able to enhance clinical quality or support the introduction of new procedures and techniques by providing novel or redesigned features. Updated equipment can also cut service delivery costs or increase the volume of work done. A business case should be able to identify these features and support the relevant equipment investment. Replacing old equipment with a new model often provides increased facilities, though not always improved cost–benefit.

### 5.4.5 Standardisation

A hospital may take out several types of medical equipment and replace them by a model – for example, switching to an update technology for blood glucose meters or lower lifetime cost infusion devices. The main drivers for this are either economic – reducing consumable and maintenance costs – or to reduce clinical risk, where a new model provides particular features or simplifies staff training, ensuring safer overall use across the organisation.

### 5.4.6 Risk and Health and Safety Issues

A clinical incident can highlight the need to invest in particular items of medical equipment. For example, a failure of decontamination may lead to a change to sterilisation systems and practices that require additional instrument purchases. New regulations may force an organisation to purchase equipment, for example, the use of improved safety containment cabinets in laboratories.

### 5.4.7 Professional and Policy-Setting Bodies

These bodies may develop guidelines for equipment replacement or usage, or with other consequences for equipment. Their content may be driven by risk considerations or be intended to push service developments towards improved patient safety, clinical quality and best practice and do not generally take financial considerations into account. The scope and urgency of what is required does, however, have to be scrutinised carefully, as healthcare organisations have to juggle priorities in meeting a wide spectrum of potential requirements and liabilities. Specific professional input is required to interpret guidelines and the organisation will need to consider what initiatives impact materially on their activities and hence which recommendations should be accepted. Ultimately, the board will need to look at the overall profile of risks facing the organisation, so that appropriate priorities can be set and timescales for acquisition planned where this is considered necessary. There may always be conflict between the high standards recommended by professional advice and policy-setting bodies and the practical limits to equipment investment that constrain the ambition of every provider organisation.

### 5.4.8 Service Developments

Service developments, which can be provided using technology completely new to the organisation, include the introduction of research techniques into clinical practice to improve patient care, or bringing in automated systems and new ways of working that cut overall costs. Extensions to existing services can be required due to increases in the numbers of patients seen or the construction of new facilities, which may include replicating facilities such as setting up a separate children's unit or delivering more one-stop or day-case care. Sometimes, an organisation may invest in high-profile and expensive technology that is unlikely to be economic but which acts as a quality improvement or marketing investment to differentiate it from other service providers.

### 5.4.9 Funding of Innovation

Research projects may develop into clinical services, based on donated or grant-purchased equipment. Apart from the clinical governance issues this raises, when non-owned equipment becomes due for renewal it can either be classified as an equipment replacement or be treated as a service development. Dealing with such cases is particularly difficult where the new treatment is neither proven nor widely accepted. Discretionary funds may be used to purchase such equipment to get round organisational controls, leaving its ongoing replacement and maintenance as a potential future burden to the organisation. Ideally, all purchase costs are included in a new development project but this is not always done thoroughly. Many healthcare building projects fail to include all the required medical equipment on their expenditure schedules, often because the necessary specialist knowledge and experience is not available at the right time.

### 5.4.10 Equipment Usage

Similar organisations can have equipment inventories that differ in size by a factor of two or more due to differences in working practices. Clinicians like to have an excess of equipment just in case a problem arises. Reviewing how well equipment is utilised and encouraging its more intensive application, for example, by setting up a comprehensive equipment library or running extended working hours, can cut down significantly the need for additional equipment and also reduce clinical risks [2].

## 5.5 Relating Funding to Need

In order to prioritise schemes and allocate major funding effectively, the organisation needs to start with reliable background information and up-to-date financial and equipment information. It must obtain support from clinical staff and managers, and this is most effectively done by having clearly defined and transparent mechanisms for bidding, prioritisation and approval of equipment investments which are applied equitably across the whole organisation. There should also be open mechanisms for managing unexpected changes in project specifications, or revised priorities, to a commensurate standard of governance. The decision-making and endorsement process, at all levels, must ensure everything is covered, from acceptance and prioritisation of local bids by directorate managers to the approval of the final capital programme by all the major contenders for capital funding, the director of operations and the board of the organisation.

Funding decisions, other than those involving minor spending, should be transparent. A department which spends a substantial amount of revenue funding on equipment should not be able to hide this when bidding for capital funds, for example, and all bidders should be open to scrutiny not only for the value of each purchase but also for the impact of each purchasing proposal on the rest of the organisation. For example, a department that takes out a managed equipment contract to bring in a type of equipment not used elsewhere in the organisation risks undermining the benefits of standardisation or volume purchasing.

### 5.5.1 General Characteristics of a System to Allocate Capital to Medical Equipment

Decisions as to what equipment to invest in should, in an ideal world, be taken in the light of what enables an organisation to best meets its objectives. Yet, in practice, other factors such as meeting legal requirements and attracting goodwill for the longer term come into play. In a healthcare context, the quality and safety of patient care is nearly always of far more significance to front-line clinical staff than financial considerations or regulations and yet the organisation has to balance all of these factors and more.

When trying to set priorities between very different types of equipment, ranging from large imaging systems to small monitoring devices, organisations face a major challenge. Is it possible to develop a central system, working to defined assessment criteria and a common framework, that ensures effective investment across the organisation whilst still taking into account local preferences and insights?

An effective capital allocation process will as far as possible keep the organisation's stock of medical equipment up to date and fit for purpose, minimise overall clinical and financial risks and provide assurance to the board and external regulators that investment is effective and meets wider governance criteria. It requires the organisation to have a good overview of its equipment stock and value and also of its strategic direction and priorities. It should be

- *Dynamic* – able to develop as the organisation's objectives change and new clinical or technological opportunities arise
- *Flexible* – can take account of rapid changes, including new funding opportunities and alterations to legislation

- *Quantitative* – able to compare options using cost–benefit criteria such as cost and income generated, and patient criteria, such as numbers of patients and degree of clinical benefit, along with risk reduction
- *Qualitative* – able to take account of local preferences and political issues around resource allocation, such as the need for long-term investment in a particular area or to keep a key clinical service at the forefront of technology or even to retain a key member of staff in the organisation
- *Realistic* – takes account of structures and systems in the organisation and the resources available to manage the process

It is rare that any healthcare organisation has enough capital to purchase all the equipment its staff might like and so any system will also need to manage end-user expectations.

### 5.5.2 Examples of Allocation Systems

The following examples are not necessarily blueprints for an organisation to follow but instead highlight issues that need to be considered when setting up a method for allocating resources. Any system has to respond to changes in organisational structures and circumstances and may be built from some or all of the components in the succeeding text, modified to suit particular circumstances. The challenge is to set up and maintain an effective system, not a perfect one:

1. *Give each management unit (such as division, department) a funding allowance*: Each unit receives an allowance to spend according to its own internally agreed priorities. Funds may be allocated proportionately to the value of existing medical equipment, calculated either by original cost or current replacement value. The advantages of this approach are its simplicity and encouragement of local decision making. It works well when an organisation's equipment needs are not changing quickly and where the emphasis is on replacement. Disadvantages are that it does not respond well to changes in technology, services or strategy which change the relative distribution of equipment value across management units. To take account of these, a separate development funding stream would need to be allocated. There is also no central oversight of the effectiveness of local decisions, which means that local allocation may be dominated by specific interest groups and not take account of the wider interests of the organisation. Without some long-term coordination of what is to be purchased, it is more difficult to standardise equipment and reduce overall costs, especially in large organisations with many different units. A possible consequence of this approach is that the organisation ends up acquiring multiple different makes and models of similar technology, which increases maintenance costs overall.

2. *Local areas bid for central funding against their top priorities*: Each ward or department prioritises its own needs and bids for equipment against centrally determined criteria. The organisation then adjudicates applications for funding and makes awards on the basis of how well bids score and what funding is available. This model assumes that bidding criteria can be determined in advance. The advantage of this approach is that it allows local knowledge to influence equipment purchasing and relative priorities whilst also allowing the organisation to set an overall strategic framework. The disadvantages are that however clear the bidding criteria might be, assessment of bids is subjective and based on partial knowledge. The more general and strategic the bidding criteria, the more difficult it is to judge between bids. Those setting local priorities have their own views on, and stake in, the outcome and may react negatively if they do not see the basis for funding allocations – so transparency is important. If the system is made mechanistic, it will not take advantage of the ability of a large organisation to favour different areas in response to changing needs and priorities and can thus ignore the potential benefits of internal brokerage between areas.

3. *Use a small number of criteria to compare the value of potential investment*: The two most obvious criteria are cost and risk. Investment return compares income generated to cost incurred; and risk reduction evaluates whether making a particular purchase will reduce clinical and financial risks. Cost-based systems weigh up costs and financial benefits and are good for ranking service development bids, if these can be costed objectively. Their disadvantage is that it can be difficult to identify full costs, especially of risk, and to value subjective benefits. Lifetime ownership costs are often much larger than the initial investment in low-value medical equipment; there will also be hidden costs, such as additional staffing or a need for extra support facilities. Total costs are therefore very difficult to estimate without good knowledge of the equipment and how it will be used. Realistic and consistent assumptions need to be made, based on expected future use. In cost–benefit analysis, it is also important to look at opportunity costs – what else might the investment be used for and can more benefit be generated that way? Benefits include the financial income to the organisation that services using the equipment will generate, and the margin of return to the organisation gives the likely payback time. Clinical benefits should be considered and where possible quantified in the widest sense, looking at the numbers of patients that will benefit from the new technology (now and in the future), what level of benefit they will receive, and any improvements that might accrue to the organisation's clinical governance. Reduction of risk (clinical, financial and organisational) should also be counted as benefit.

4. *Risk reduction*: This is linked ideally to an organisation's wider risk register. Its evaluation needs to consider the starting level of risk and also reductions any investment will create, with a clear system for estimating risk so that bids can be compared to each other. One method is to allocate equipment bids into risk rankings. Top priority is given to ensuring clinical services continue and to complying with legal requirements, with lower priority for bids addressing improvements in service efficiency and quality. The advantages of a risk-based approach are that it is ideally suited for ranking issues of health and safety and risk and also works reasonably well when looking at equipment replacement. Disadvantages are that it is difficult to use to take account of the return on a financial investment, as part of a technical or service development.

5. *A hybrid system that takes account of multiple variables*: Such a system tries to take into account local priorities, risk issues, cost and benefit, organisational developments and strategies and reflect patterns of equipment ownership. There is no simple mechanistic way of doing this and any process needs the application of informed judgement. It should also aim to use as much knowledge as possible about what is happening locally and to link this to an organisation-wide view. The advantage of a hybrid system is that an organisation can tailor what it does to meet local requirements based on the size of the organisation and local priorities. However, such systems are more complex to understand and to implement.

6. *Set up a capital replacement programme*: A capital replacement programme is an important component in setting up an effective equipment funding system. Significant work is needed to keep equipment up to date in a large organisation, even just major items. It relies on having an up-to-date medical equipment inventory and capital asset database that is reviewed each year in the light of what has been invested and whether clinical needs have changed. Comparing what equipment has been bid for to what is actually bought, and whether it is being used as expected, is a useful exercise to check how different groups approach the introduction of new technology and whether they are effective in predicting the effects of changing clinical practice and technology on future equipment needs. Replacement programmes are useful points of reference, to help predict investment required over following years and ensure equipment is replaced in a timely manner. A possible approach for a major hospital is to map the detailed replacement of items individually costing over £100,000 and of grouped items like haemodialysis units costing over £250,000, whilst making an annual aggregated allowance for replacing smaller items.

### 5.5.3  Creating an Effective System

The mechanisms outlined earlier need to be taken into account when an allocation system is devised. They must be complemented by introducing elements of fairness and governance that make the system politically and emotionally acceptable to all the participants. Hence, in any allocation system, clinical and managerial involvement in decisions is essential. Including clinical staff encourages them to articulate their needs better and use capital more effectively, whilst senior managers bring a strategic and financial view. The best decisions come from open and honest debate and challenge between these different perspectives in a clear and transparent processes that keep all those involved informed of progress and has the ability to take both local and institutional priorities into account.

#### 5.5.3.1  Flexibility

The system must be able to provide for the long-term and yet be able to respond flexibly to sudden changes. Hence, there must be a long-term plan to replace both expensive items and large groups of less expensive equipment, kept updated from year to year as equipment is purchased and service priorities change, together with a clear understanding of the funding available to the organisation and the levels of risk it is willing to bear. There must also be an ability to incorporate service development projects. Short-term adaptability will require regular review of equipment needs through the year and a contingency allocation to deal with equipment failures and other emergencies. A defined proportion of funding may be kept back as a reserve, or money redirected from projects that are abandoned or held over to another year. Another approach is to have a list of second priority projects that will go ahead if any of the reserve fund is left over.

#### 5.5.3.2  Taking Organisational Politics into Account

The political nature of resource allocation cannot be ignored. However logical it might be, an allocation process which is too rigid to take account of varying opinions and provide for a degree of negotiation and real involvement in decision making will not be popular and individuals are likely to seek ways to circumvent and undermine it. Giving local users a chance to state their priorities and respecting these is important, as end users may well be aware of factors important to their services that are not articulated easily in formal application forms. Above all, the process must be overseen by an individual director who carries the confidence of both clinical and managerial staff, such as the medical director. The outcome of the process will be a prioritised list of investments that have been reviewed and accepted by clinical users or their representatives, together with an understanding of the risks and opportunities these represent. The ultimate decision on funding will be taken by the board, in the light of available finance and its view of associated risks.

### 5.5.3.3 Setting Realistic Time Scales

The time and effort required to purchase equipment should not be underestimated. Good practice means investing time to identify requirements, drawing up clear clinical and technical specifications, examining options and weighing the advantages and disadvantages of different makes and models of equipment. Larger projects require a proportionately greater investment of time and may need a major project to integrate equipment installation and building works. Lead times to installation of 1–2 years are not unusual for projects costing many hundreds of thousands of pounds, such as installations of radiotherapy and major imaging equipment.

### 5.5.3.4 Gaming

Gaming of the allocation system is inevitable and a good system should provide ways to identify and correct for it. For example, departments may give a service development bid a high priority but assign a lower priority to a piece of equipment that has failed, knowing that the latter is likely to be replaced anyway. Another example is when equipment is assigned a higher risk score than it warrants by scoring the highest consequence with the highest likelihood, even though there are contingency arrangements in place to limit any risks, rather than scoring the realistic likelihood of the highest consequence. For example, a piece of equipment may fail sometimes, and for most of the time this may simply be a nuisance. However, it may occasionally fail in a way which puts patients or services at risk. The correct risk scoring is either a likely occurrence of minor service disruption (a less severe consequence) or the unlikely occurrence of a serious incident.

Another way to identify gaming is to look in detail at the history and context of each bid and the clinical need it addresses. For example, requests to retain old equipment just in case need to be looked at closely, as a department may keep an item replaced in earlier years as a spare and then put it forward again for replacement in a later year, effectively trying to replace it twice. Items which were funded originally from research or charitable sources may come up for replacement without the underlying clinical service necessarily being approved by the organisation. It is therefore essential to review all bids with both the medical director and the operations director to see if the clinical need, and the continuation or expansion of the related service, is justified.

## 5.6 Outline of a Possible Bidding Process

The aim of the bidding process is to allocate as effectively as possible the organisation's funds for medical equipment purchase. We outline an example process here that is based on a hybrid scheme using risk-based scoring to prioritise clinical risk and financial measures to commit to service developments. Figure 5.1

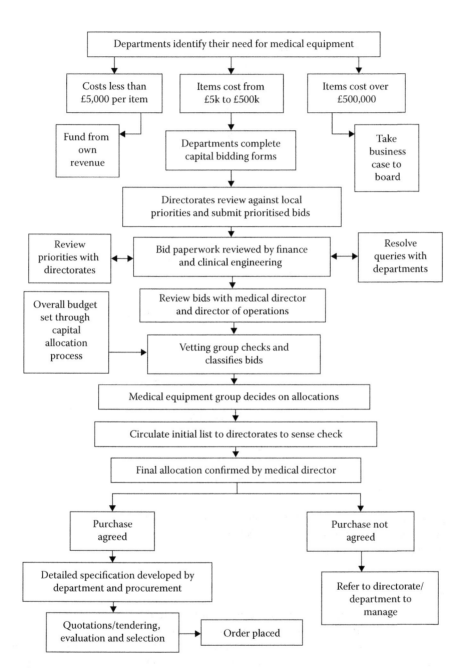

**FIGURE 5.1**
Possible medical equipment bidding process.

**TABLE 5.1**

Factors to Consider When Setting Purchasing Priorities

| Main Driver for Purchase | Factors to Consider When Writing and Reviewing Bids |
|---|---|
| Health and safety/clinical governance/statutory need | • Extent of risk to patient safety and to the trust<br>• Impact on trust operation if not funded<br>• Legal and liability exposure<br>• Meeting regulatory requirements and other priority targets |
| Service/quality/ productivity | • Level of reduction of clinical risk<br>• Impact on patient care<br>• Level of improvement in quality<br>• Income generation and costs<br>• Priority against government and local clinical targets<br>• Progress towards corporate objectives |
| Routine replacement | • Proportion and size of inventory requiring replacement<br>• Level of replacement backlog against expected life<br>• Clinical need for the service |
| Equipment failure and/or reliability | • How often equipment is used and on how many patients<br>• How critical it is to a service and to trust objectives<br>• Consequences of and contingencies to cover failure<br>• Frequency/severity of equipment problems, particularly if supported by an objective technical assessment |

**TABLE 5.2**

Scoring the Likelihood of an Event Occurring

| Risk Likelihood Score – Probability/Frequency | | |
|---|---|---|
| 5 | Almost certain | Will probably occur frequently |
| 4 | Likely | Will probably occur, but not as a persistent issue |
| 3 | Possible | May occur |
| 2 | Unlikely | Not expected to occur |
| 1 | Rare | Would only occur in exceptional circumstances |

presents a flow diagram of the process and Table 5.1 the major factors considered in scoring. Tables 5.2 through 5.4 set out a risk scoring scheme for classifying bids into risk categories.

### 5.6.1 Writing Bids

Departments complete a request form for each item being bid for. The proforma asks for background information on existing items, a clinical justification of need, identification of costs, a description of the item(s) sought and scoring of the major risks from not purchasing the item. An end user, clinician or manager who may have originated the project will explain the need from their detailed local knowledge, whilst a clinical director or senior manager will endorse the requirement as appropriate. Directorate managers

**TABLE 5.3**

Scoring the Consequence of an Event Occurring[a]

| | Risk Consequence Score by Category | | | |
| --- | --- | --- | --- | --- |
| | Injury/Harm | Service Delivery | Financial | Reputation/ Publicity |
| 5 Catastrophe | Unanticipated death/large number injured or affected (e.g., breast screening errors) | Breakdown/ closure of a critical service | £5M | Long-term/ repeated adverse national publicity undermines patient and/or referrer confidence Chair/CEO removal |
| 4 Major | Major permanent loss of function for patient unrelated to natural course of illness/ underlying condition/ pregnancy/ childbirth | Intermittent failures in a critical service | £1M–£5M | Widespread and sustained adverse publicity Increased political/public scrutiny |
| 3 Moderate | Semi-permanent harm (1 month–1 year), >1 month's absence from work for staff | Sustained period of disruption to services | £100k–£1M | Widespread or high-profile adverse publicity |
| 2 Minor | Short-term injury (<1 month), 3 days to 1 month absence for staff | Short disruption to services that affects patient care | £10k–£100k | Adverse publicity |
| 1 Insignificant | Minor harm Injury resulting in <3 days absence from work for staff | Service disruption that does not affect patient care | <£10k | None |

[a] Where risks have consequences in more than one column, use the score from the highest column.

are then asked to put items from their own unit into a single priority ranking, as this helps to inform allocation decisions by indicating clinical and service development priorities. Equipment which was not purchased the previous year has to be resubmitted for consideration, as otherwise it would not be clear if it was still required.

Many medical equipment bids are not well written and omit critical information. Clinical users need to be prepared to explain their needs simply and straightforwardly in simple and unambiguous language and to take the trouble to fully account for lifetime equipment costs including maintenance, staffing and consumables. This is particularly important for large bids, where

**TABLE 5.4**

Grading the Risk

| Consequences → Likelihood ↓ | None 1 | Minor 2 | Moderate 3 | Major 4 | Catastrophic 5 |
|---|---|---|---|---|---|
| 5 Almost certain | Green | Orange | Red | Red | Red |
| 4 Likely | Green | Orange | Orange | Red | Red |
| 3 Possible | Green | Green | Orange | Orange | Red |
| 2 Unlikely | Green | Green | Green | Orange | Orange |
| 1 Rare | Green | Green | Green | Green | Green |

*Note:* Green is low risk, orange is moderate and red is high.

unidentified items and poor assumptions can lead to large underestimates in the cost of a project which only become apparent when it is too late to cancel a purchase. Bids have a better chance of being funded if accompanied by a detailed case and supporting documentation. Where possible, bids should be put in the context of ongoing equipment replacement and other developments, such as a programme of work to meet higher service workloads or tighter clinical standards.

### 5.6.2 Ranking Service Development Bids

It is difficult to rank the merits of service development bids against equipment replacement bids when using a standard risk matrix. However, financial elements are more straightforward for development bids, particularly where there is a short payback period. Like any business case, it is important to ensure that all relevant costs are included. Information submitted with equipment bids is often too limited to be able to assess the full costs and value of individual developments, even where the user is clear that the bid is cost-effective and/or will bring a marked improvement in the quality of patient care. All service development bids involving the purchase of capital equipment should be notified through the capital bidding process, whether they make savings within a management unit or improve care across the organisation.

Development bids should be identified as a separate category and be scrutinised separately, with more detailed business cases being produced as necessary to obtain organisational support. Consideration should be given to prioritising service developments separately to equipment replacement, with the aim of identifying a pot of money to invest in service and cost improvements. Larger entrepreneurial bids will need to have a business case taken to the finance and medical directors and the board to ensure all relevant factors are evaluated. They will need to fit with longer-term service strategy, provide cost–benefit and address research, teaching and the introduction of new technology, as appropriate, whilst having a manageable impact on other areas. Projects requiring external funding will undergo additional scrutiny before being put forward, to make sure they are an organisational priority and are sustainable. External scrutiny from outside funding bodies is also likely to be challenging. Financial limitations put such a bid in competition with other calls on capital, however good a case it may have, unless there is ready access to external funding.

### 5.6.3 Bid Vetting

Decisions as to which bids are accepted will probably take place at very few meetings, and perhaps only one. These meetings need high-quality information and bids that are well presented and coherent. Having all major stakeholders present – unit managers, lead clinicians, etc. – can make discussions very lengthy and contentious, particularly where individuals regard themselves as being present to push through bids from their own area. Working from an initial prioritised list makes it easier for the meeting to consider which bids are ranked correctly and where the cut-off point of funding should lie, and an important part of preparation for the meeting is to produce clear, short summaries a couple of sentences long highlighting the key features of each bid. Some time before this definitive meeting, therefore, a small group with representation from finance, procurement, clinical engineering and operations meet to scrutinise the original bids, prepare suitable summaries and establish initial priorities. Group members check that bids classed as capital actually qualify and identify any bids that might be combined. They confirm whether projects are correctly classified as replacement or service development bids and check for critical gaps or anomalies in the information presented. Finance representatives review the figures and any projections and costings. Procurement looks at historical pricing, standardisation, consumable consequences and procurement concerns. Clinical engineers comment on the history and current state of equipment, looking at records of breakdown frequency, and consider other technical options and developing technologies. Any uncertainties are clarified with bidders. Group members check the realism of risk scorings and review them against risk registers, adopting a standardised approach to scoring and altering original risk estimates where appropriate. If the bid is for a large project that might impact on supporting clinical areas or other services such as IT, estates, radiology or decontamination, the group flags this up to check that any demands can be met. The group may also suggest alternative funding sources. Finally, contentious or major bids will be checked in detail with the operations director, to see whether they support the organisation's strategy and business plan. Based on all this information, the group produces an initial prioritisation which is agreed with the medical director. First priorities are set using risk and financial criteria and knowledge of the likely financial envelope. Spreadsheets and other summaries are prepared showing the major relevant issues for each bid.

### 5.6.4  Decision-Making Process

The effectiveness of the allocation process depends critically on how the decision-making group operates and how well information is relayed to it. It is remarkably difficult to explain complex operational issues in a succinct way that makes it easy to compare different bids, let alone decide on the merits of one clinical requirement over another. The structure of the decision-making group is critical. Whilst the medical equipment management committee (see Chapter 4) can facilitate the decision-making process for project approval and may review overall risks, it is unlikely to have the correct clinical and director level representation to allocate large blocks of capital effectively or weigh up the overall value of one investment against another. A second group – the medical equipment group – may be convened to oversee annual allocation of capital and contingency funding, as well as review urgent unplanned equipment demands in the light of project delays, cancellations and funding variations during the financial year.

### 5.6.5  Confirmation

Once the decision-making group agrees on priorities, a list of bids to be funded is circulated for comment before being passed to the board via the finance director for final sign off. Approval through the capital bidding process is only for funding, not for a specific make or model of equipment. Standard procurement procedures must be followed, based on end-user specifications, appropriate tendering and the evaluation of equipment from different suppliers, as discussed in Chapter 6. Decisions on whether items are purchased outright or leased may be taken by the capital finance function but should not affect medical equipment purchasing priorities.

### 5.6.6  Procurement

The purchasing process only starts once bids have been approved. This is to ensure that effort is not wasted and that the trust gets best value for money. Bidders should base costings on guide prices and not get into negotiations with manufacturers or seek detailed quotations, as this then gives suppliers an insight into the possible funding that the organisation will put aside to purchase a particular item. Large capital projects usually require formal tendering, due to national regulations and standing financial instructions, and it is a waste of resources for both the organisation and the manufacturer to go to tender before a bid is approved and have it subsequently fail. Initial estimates will therefore differ from final spending. Where final quotes differ substantially from the nominal budgeted by a substantial amount, the decision-making body must be alerted so that the agreed purchasing programme is not disrupted. Substantial variations from the original bid due to issues other than finance, such as supplier choice, technological changes or extended specifications, will also need to be approved. As well as overspends, some projects may be delayed or fail altogether resulting in underspends. The cumulative effects of these variations will affect the total available, so chasing up expenditure needs to be done actively and regularly, with bids traced through the process and with clear links back to finance. There must be ongoing review of overall expenditure and equipment priority as the financial year progresses. Cut-off dates should be set for process milestones, particularly regarding holding over expenditure from one financial year to another. Very large projects may need to be split over more than 1 year. Chapter 6 goes into procurement processes in more detail.

### 5.6.7 Equipment Replacement outside the Annual Capital Allocation Process

Items inevitably come up for purchase during the financial year, due to unplanned equipment failures and service development requests. It is difficult to put these items through exactly the same vetting and prioritisation process as those coming up in an annual bidding round, even if resources are set aside for this, as by their nature these demands often require urgent approval. Requests for in-year equipment purchases should be submitted on the same documentation as for the annual bidding round. A good compromise between measured consideration and a fast response can be achieved by combining further meetings of the medical equipment group through the year to review incoming bids, perhaps every 3–4 months, together with authorisation by the chair of the group (usually the medical director) to approve urgent bids. Even where matters are urgent, suitable expert comment will need to be provided to the medical director to ensure critical issues are considered.

## 5.7 Summary

There is never enough money to meet everyone's wishes for new equipment, not only because resources are limited but also because new and improved equipment constantly appears on the market. Decision makers need to find ways of stretching and supplementing available resources as far as possible and we considered ways of doing this using capital, revenue, charity and research funding and by leasing or ways to avoid outright purchase. It is a difficult and complex process to satisfy everybody that a fair allocation between the conflicting demands of routine clinical services, service developments and research applications has been reached. There is no best way of making decisions and all organisations struggle to achieve the best value they can from equipment investment. In this chapter we looked at the characteristics of possible systems and outlined a process that would incorporate beneficial aspects from all of them.

## References

1. Auditor General. 2003. Managing medical equipment in public hospitals. Government printer for the state of Victoria. http://download.audit.vic.gov.au/files/medical_report.pdf (accessed August 23, 2013).
2. Quinn, C., Stevenson, E. and Glenister, H. NPSA infusion device toolkit: A cost-saving way to improve patient safety. *Clin. Govern.*, 9(3), 195–199, 2004.

# 6

## Procurement, Specification and Evaluation

### 6.1 Introduction

Technology, and its contribution to the activities of healthcare organisations, is changing ever faster. Like-for-like replacement of a worn out item of hospital equipment may not even be possible as the last model is unlikely to be the same as the current one or meet the latest clinical need. Updated diagnostic or therapeutic technology can provide improved and more efficient patient care. Demand for common devices such as basic patient monitors is changing in developed economies as the number of hospital beds reduces and the amount of sophisticated monitoring increases. Telemetry and central monitoring are developing hand in hand with hospital at home facilities and a demand for robust, network-linked devices in primary care. This has practical implications when setting up new services, as the equipment eventually needed may vary significantly from that envisaged in the early planning stages of, say, a new hospital. When acquiring new or replacement equipment, therefore, it is wise to consider different approaches to service provision and the possibility of adopting new technologies or approaches. Maintaining flexibility is also a good investment in the longer term.

This chapter aims to provide an overview of the procurement process and an understanding of what is involved in evaluating new equipment for purchase, from single items to the wholesale adoption of innovative technologies. It includes an overview of the role of clinical engineering in technology procurement and complements material introduced in Chapter 2.

### 6.2 Approaching a Replacement Programme and Tender

Before directly replacing any item, the need for it should be evaluated from three perspectives (Figure 6.1). The first question is strategic: 'Does this service need to continue?' The answer comes from senior management and will depend on the organisation's overall strategy and how it perceives the

**FIGURE 6.1**
General factors to consider when deciding whether to purchase equipment.

financial income or other benefits it will get from a service. The second perspective comes from the clinical service: 'How is current technology being used and is there a better way in which this service can be delivered?' Both finance and quality issues need to be taken into account, and it is helpful to look outside the organisation to see whether innovative solutions have been developed elsewhere and whether they can be adopted in-house. The third view is at equipment level: 'What is the most appropriate technology to meet this clinical need?' Simpler equipment may match a service specification and be cheaper but a different and more expensive technology might reduce the number of steps in a process and improve overall efficiency. A business case is a formal way to present these issues.

In-depth evaluation is not required for every item but is worth doing for models where significant numbers are used or where the overall financial investment is substantial. A replacement decision might take into account whether or not equipment is in current use, what risks are presented by not replacing it, whether the clinical need it satisfies could be met using existing items without a reduction in service and whether items transferred from another area might be suitable. Once the need for equipment has been demonstrated and funding identified on the basis of a nominal device-type description (see Chapter 5), the procurement process starts with some type of tendering based on detailed technical and functional specifications, followed by evaluation of tender responses and final selection of the device.

For most projects where large capital equipment or fleets of equipment are bought outright, the overall value will be high enough to require tendering under national or institutional rules or to follow good practice. Under some circumstances, tender waivers can be used, most commonly on the grounds that only one model fulfils the desired clinical function – particularly relevant with innovative technology – or to ensure new equipment is compatible with existing items. Also, pre-tendering exercises may be carried out by institutions, government bodies or appointed agents allowing an individual organisation to choose certain models without issuing a specific tender. Taken together, these three potential tender exemptions for uniqueness, compatibility and pre-tendering allow a sizeable proportion of a hospital's equipment purchases to bypass individual tenders. Even so, it is good practice in many cases to draw up an equipment specification and evaluate equipment against it even if a tender is not being undertaken, as this will clarify the purchaser's requirements and the supplier's understanding of what is required. It also provides useful evidence for any legal process if equipment does not deliver the promised performance. Where equipment has been pre-tendered nationally, it may still be advantageous to go to tender in an attempt to obtain lower prices or other concessions.

In many cases, the tendering process is slow when compared to purchasing off the shelf equipment with direct quotations. For example, the European public sector tender process is necessary for purchases costing approximately £100,000 and over. The full process takes at least 77 days from start to finish. After this, it typically takes weeks to months to evaluate tenders and carry out clinical and technical trials on different models. Tendering thus consumes considerable time and effort but is much more than an irksome chore to be avoided wherever possible. It can ensure that an organisation will have suitable and reliable equipment for years to come and offers positive benefits which include the following:

- *Finding new suppliers*: New companies may have entered an established market and be able to compete effectively on reduced price or other benefits.

- *Gaining additional benefits*: Previously optional functions may be standard on new models. This can reduce the need for separate stand-alone devices.

- *Taking advantage of the latest prices*: Equipment with enhanced functionality might be available at a cost similar to established models.

- *Improved market intelligence*: Increased interaction with more suppliers contributes to background knowledge of what is on offer and from whom.

- *Enhanced governance*: Tendering helps the organisation hold the manufacturer to account if equipment does not meet its stated purpose or specification.

- *Support for standardisation*: A multi-year framework agreement can be set up with substantial discounts on the basis of a large number of eventual purchases.

Where relatively small additions or replacements to existing fleets of equipment are required, the pros and cons of evaluation and tendering should be balanced against the consequences of simply reordering the type in use. This tension between standardisation, cost and good corporate governance presents dangers, as illustrated in the following two examples.

**Example 6.1   Inertia by Stealth**

A hospital uses a large fleet of patient monitors, consisting of a single model. Consumables, user training, maintenance and support arrangements are in place and working well. A single item then fails and needs replacement. It makes operational sense to replace it by the same model. Modest increases to the number in the fleet may be made using the same arguments, perhaps over several years. Eventually, the model becomes obsolete. Meanwhile, older units begin to fail and require replacement. The manufacturer introduces a new model which has much in common with the old: compatible consumables, exchangeable modules and interconnectivity and similar control software. It makes sense to use the same manufacturer as support and training is easier and clinical users like continuity. Over time, this situation continues by default and the hospital never market tests its supply of monitors, in direct contradiction to the ideals of procurement. Once a model becomes widespread and standardised throughout an organisation, the argument for replacing it by anything else becomes more difficult to justify.

**Example 6.2   It's the Rules**

An organisation can waste much time and effort by rigidly following a tendering process for even minor extensions or replacements. If a purchaser hops between suppliers to gain short-term price or other advantages, the benefits of standardisation are lost with increased risk and costs in the long term. If this process results in the existing model being chosen in any case, the tendering process has been wasteful for both purchaser and supplier.

A pragmatic solution to the conflict between standardisation and innovation is to maintain a register of standard models which departments must order from, and to review the register at appropriate intervals. For each category of device (patient monitor, syringe driver, anaesthetic machine and so on), there will be a standard model bought as the need arises unless there are special requirements that dictate otherwise. Standard models should be selected on the basis of evaluation or competitive tendering, and be updated when a large-scale fleet replacement is needed, when there is a major advance in technology, or at an agreed default interval such as every 5 years. Standardisation can therefore combine economies of scale with flexibility to purchase when required and achieve discounts for bulk purchasing of devices, accessories, consumables, spares and further training. Indirect costs are further reduced by eliminating confusion between device models, which reduces risk and the rate of user error. Service uptime is increased by an ability to substitute whole units, subassemblies and accessories in the event of device failure, and user and in-house maintenance is simplified by familiarity with a particular model. There is thus conflict between the advantages of standardisation in keeping the complexity and costs of procurement to a minimum and a desire to keep up with technological advances and provide an even-handed assessment of available models. There is also a balance between reducing the likelihood of user errors and increasing the potential for exposure to financial risks in the event of problems with a particular model, and the organisation must weigh up all these factors when considering what to do.

Effort spent on specification and procurement for minor projects is proportionally much lower than for a full tender. If a small number of devices of a particular type is required and their cost is low, the process can be as simple as obtaining quotations from two or more suppliers and choosing the equipment by a process involving key individuals. The decision-making process should still be documented, however, and be transparent in case of supplier complaint or auditor investigation.

Once a decision has been taken to tender, the outline process is as follows:

- *Call for expressions of interest*, specifying numbers and types of device required and details of how responses will be evaluated.
- *Evaluation of initial responses*, producing list of suppliers to be invited to tender. In Europe, this is done by means of a pre-qualification questionnaire.
- *An invitation to tender* is published, whereby suppliers are given detailed technical and performance specifications for the required devices.
- *Suppliers produce tenders* with details of how their products meet the specification, with costs and details of any support, consumables and maintenance on offer.
- *Evaluation of tenders* by financial and technical desktop examination.
- *Shortlisting of tenders.*
- *Evaluation of the physical device* by clinical users and technical support staff, with testing, as appropriate.
- *Selection of the most suitable supplier*, followed by detailed negotiations and purchase.

An example of tendering and the processes involved can be found by looking at European procurement rules [1] and UK regulation [2]. These are also relevant outside Europe as the Government Procurement Agreement, an international agreement negotiated by the World Trade Organization, gives suppliers from a number of countries including Japan, China and the United States the same rights as EU suppliers [3]. European regulations aim to open up public procurement in the EU to the free movement of supplies, services and works and end discrimination on the grounds of nationality. They require public procurement to be based on value for money, defined as 'the optimum combination of whole-life cost and quality to meet the user's requirement'.

## 6.3 Preparing a Specification

Both the organisation and its potential suppliers need to understand what its exact needs are, so that any devices purchased are fit for purpose. It needs to specify the clinical and technical performance it is looking for and to identify which requirements must be met in full and which ones are desirable. Once product evaluation starts, the specification should be fixed but there is scope to make minor variations as long as all potential suppliers are informed. It is the supplier's performance against the final specification, together with costing and other financial information, that forms the basis for making a decision on what equipment to purchase.

Internal governance issues may also need to be dealt with when developing a specification and carrying out evaluations. Although technical, clinical and cost issues are considered objective by nature, some users have a determined preference for or against a particular company's products or want to play the system by putting down vague specifications that allow the purchase of additional unspecified equipment after funds have been released. In Chapter 2, we discussed how clear specifications can be a safeguard against prejudice and undesirable practices and stop individuals bypassing a fair and open process. Here, we are more concerned with the details of specification as an aid to ensuring technical and clinical performance.

European public procurement regulations set out criteria designed to ensure that all suppliers in countries covered by the rules are treated on equal terms. Specifications should set out performance requirements but must avoid specifying how a device works. This is because a technology may be proprietary to one manufacturer, whilst others may provide an equivalent function through a different design. Since this process is intended to eliminate discrimination, the real-life test of a specification is whether it gives sufficient grounds to provoke complaint by an unsuccessful bidder. Governance processes, including scrutiny by auditors and the expertise of clinical engineers and procurement specialists, should warn healthcare organisations if they are in danger of running into trouble in this area.

Device specifications should set out any particular need for compliance with legal requirements, standards, guidelines and import licensing conditions. They should also state any local and international technical standards that apply, including the provision of safety features and recording of exposure in radiological imaging equipment. Particular requirements of the medical device regulations in most countries will be known to suppliers and requests for information are routinely incorporated into procurement procedures. The standardised pre-purchase questionnaire (PPQ) used in the United Kingdom is intended for use when purchasing electrical medical, dental, ophthalmic and laboratory equipment and asks for essential information such as compliance with regulations, CE marking, maintenance arrangements and training support. In addition to the generic form, some product-specific versions have also been developed.

Most specifications do not need to be drawn up from scratch. It is often possible to build on previous specifications covering similar devices and to modify them to suit a particular situation. Standard templates and associated guidance are produced by some professional organisations. Two examples are the documents designed by the UK Royal College of Radiologists [4] to assist with ultrasound scanner specification and a questionnaire produced by the British Nuclear Medicine Society in conjunction with the Institute of Physics and Engineering in Medicine for gamma camera purchase [5]. Specifications are usually drafted by one or two people but are overseen by a multi-disciplinary team involving all those who will use and support the device. It is impossible here to cover even a fraction of the different technical and other elements that could be included in detailed specifications across the full range of device types available. What is important is that those preparing specifications should be experts in the relevant clinical and technical fields and appreciate the subtle details that differentiate potential solutions. Basic information such as the required quantity and delivery deadline should also be included in the tender and, if relevant, should ask for the price of additional units to be held for a defined period.

A purchaser should ask for a tender response covering the costs of ancillary items such as consumables and accessories, to be able to assess lifetime running costs and the possibilities of volume discounts. The same applies to details and costs of the available maintenance support from the supplier, including the potential for in-house maintenance partnerships where appropriate. Some useful questions are whether or not routine software updates are included in the purchase; the costs of major spares such as x-ray tubes, ultrasound transducers and circuit boards; the availability and costs of maintenance training for in-house maintenance staff; and access to software and hardware maintenance tools. Chapters 8 and 9 provide a more detailed consideration of issues relating to maintenance provision and service contracts. Decontamination methods and processes are specified by manufacturers but may not be readily available to the purchaser, so these need to be clarified. Since training in the use of devices is essential for end users (see Chapter 7), it is important to find out what training is available from the supplier, how it is delivered, and at what cost. The organisation must consider whether it can provide future training in-house, using manufacturer or other support materials, as otherwise it will have to rely on continued manufacturer support and must cost for this.

A purchaser must also consider infrastructure requirements, particularly for larger items of equipment. Decisions need to be made as to whether changes to space, power supplies, cooling facilities and so on are required to accommodate new or replacement equipment. The extent, feasibility and cost of potential works may limit the range of equipment that can be afforded and so the equipment tendering process must analyse tender responses to identify any underlying infrastructure support requirements. New buildings should provide for the future by creating facilities capable of taking equipment from different manufacturers and potentially allowing for future technical developments. A choice also needs to be made as to whether infrastructure works are specified and project managed by the healthcare organisation, with the actual works carried out in-house or contracted out, or whether the whole project is outsourced to a third party. It is possible to ask the manufacturer to provide *turnkey* project management whereby they plan and commission all building and infrastructure works alongside supplying the equipment. This solution is often adopted for large capital installations such as imaging equipment, where there are significant specialist requirements. Where a hospital manages its own equipment in association with a private finance initiative (PFI) infrastructure project, the PFI provider may insist on using its own builders and project managers to carry out any works unless this is excluded by the contract (see Chapter 4). Whatever method is used to execute a combined equipment and infrastructure development, the healthcare organisation should commission its own specialist advice. If an organisation does not have staff to do this, it must purchase advice from qualified experts who can challenge the actions of a third-party supplier, through a legal process if necessary. Every effort should be made to resolve conflict before a project goes ahead and to incorporate detailed requirements into contracts with the supplier. Infrastructure upgrades may be subject to a separate tender or be included as part of the overall equipment purchase.

---

**Example 6.3   Physiological Monitoring: Functional Specification**

Rather than set out a number of abstract points on specification, we consider one example – equipping a hospital with physiological monitors – which illustrates some of the major issues involved.

Physiological functions to be monitored might range from a simple ward system measuring ECG, non-invasive blood pressure (NIBP) and arterial oxygen saturation ($SPO_2$) to a full system for the intensive care unit covering ECG, invasive blood pressure, NIBP, temperature, $SPO_2$ and respiratory $CO_2$ and $O_2$ measurements and incorporating other measurement and analysis functions.

The organisation will need to decide whether to use a single standard model for all purposes or whether to have two or more models of differing complexity. In the first case, a standard model can comprise a core unit for wards that provides basic functionality but to which extra modules and software can be added to create a full function monitor suitable for intensive care monitoring. The alternative is to purchase a basic model for the wards and a different, more sophisticated one for intensive care. The flexibility and advantages of complete standardisation must be weighed against the likely costs and relative numbers of each type of monitor required.

Another monitoring requirement is for small portable machines to act as transport monitors to support patient movement between theatres and intensive care and to cover ambulatory monitoring. Here, small size, lightness and portability may be key requirements. There is a trade-off between robustness, usability and weight, with factors such as the expected time of operation on battery power, charging time and the ability to survive being dropped, being important factors in any specification. Environmental robustness is important for portable equipment intended to be used outdoors, for example, in hot, cold, humid or damp conditions. Robustness with power supply variations, surges and outages may be important and units may need uninterruptable power supplies, appropriate battery capacity or non-volatile memory. Portable monitors should ideally have some degree of compatibility with other monitoring on-site, to be able to store and transfer data so it can be accessed as the patient moves round the organisation, usually in a transferable memory module. Some monitors, particularly transport monitors, may be used in hazardous zones such as magnetic resonance imaging (MRI)-controlled areas and must be of specialist design for safety reasons. Others may require specialised functions such as ECG gating for cardiac nuclear medicine applications.

Other necessary decisions include whether central monitoring is required on some wards, whether telemetry is needed for areas such as side wards and for ambulatory patients and whether or not a networked central data gathering system should be installed to record and analyse electronic and manual inputs for clinical management, costing and research purposes. Detailed fault and event logs in intensive care and transport monitors provide traceability if there is an adverse incident. Display screen sizes for the various clinical areas must be specified alongside local technical details such as whether ECG monitoring is to use 3 or 5 leads and whether chart print outs are needed in some clinical areas.

Questionnaires and checklists based on the specification are a useful way to structure evaluations and capture results. Pre-existing examples can guide their content, and clinical and technical representatives with a reasonable level of experience and expertise should be involved, along with specialist scientific staff and safety advisers. Forms need to be completed and evaluated objectively, with cross-checking between individual assessors to make sure there is no individual bias. For both clinical evaluation and technical evaluation, it is best to use forms with a structure that provides for quantitative scoring and free field comment. Generic examples of a form for clinical and technical evaluations are given in Tables 6.1 and 6.2, respectively.

**TABLE 6.1**

Schematic Technical Evaluation Form

| Device | Clarity of Technical Manuals | Ease of Decontamination | Consumables and Spares Price and Availability | Maintenance Ease and Cost | Robustness and Reliability | Performance Against Specification | Special Issues | Comments |
|--------|---|---|---|---|---|---|---|---|
| A | | | | | | | | |
| B | | | | | | | | |
| C | | | | | | | | |
| D | | | | | | | | |
| E | | | | | | | | |

*Note:* Give marks 1–5 for each device for each assessment, where 1 = very poor, 2 = poor, 3 = acceptable, 4 = good and 5 = very good.

**TABLE 6.2**

Schematic Clinical User Evaluation Form

| Device | Clarity of Instructions | Ease of Use | Performance in Service | Physical Robustness | Display/ Controls | Consumables Ease of Use | Special Requirements | Comments |
|--------|---|---|---|---|---|---|---|---|
| A | | | | | | | | |
| B | | | | | | | | |
| C | | | | | | | | |
| D | | | | | | | | |
| E | | | | | | | | |

*Note:* Give marks 1–5 for each device for each assessment, where 1 = very poor, 2 = poor, 3 = acceptable, 4 = good and 5 = very good.

## 6.4 Tender Receipt, Evaluation and Decision

### 6.4.1 Initial Process

Where tendering comprises a two-stage process, as in the European restricted procedure [1], shortlisting takes place after evaluating expressions of interest. Shortlisted suppliers are invited to tender to the full specification and a closing date for receipt of tenders will be set. Companies should be allowed time to contact the purchaser to clarify issues in the tender and the purchaser should copy responses to all shortlisted suppliers. Tenders should be sealed on receipt and must only be opened once the tender deadline has passed at a single session where at least two authorised people of sufficient seniority, such as the heads of procurement and clinical engineering, record and cross-check salient features such as major cost breakdowns on a tender summary sheet. The aim is to provide transparency and make it difficult for one person to discriminate against a company by distorting information. Tender documents including any relevant correspondence and notes need to be maintained for a certain period in case of any appeal against a tender outcome.

Once the range of likely equipment on offer has been established, more detailed evaluations can begin. There are three types of evaluation: financial, clinical and technical, carried out by groups which include procurement experts, clinical users and technical specialists. The first looks in detail at suppliers and their track record and at relative tender costs and offerings. The second involves end users and experts in the clinical area concerned, to see how easy the device is to use and how well it meets clinical need. Finally, a technical evaluation by engineering and scientific experts aims to establish how well the device performs, what limitations it might have and whether it is compatible with existing equipment and processes. Questionnaires and checklists are very helpful for keeping formal records and setting out comparisons.

There is no substitute for direct hands-on experience when trying to establish ease of operation and use, and suppliers should be approached to loan equipment or provide access to it elsewhere. Examining, operating and testing equipment allows potential users to verify whether a specification is met or if a device requires an add-on at additional cost. The group planning the evaluation should work together to review and approve each of the evaluation topics, questionnaires and checklists so there is consistency and adequate coverage of all areas. In practice, every evaluation must be tailored to the clinical requirements and operational environment pertaining at the time.

In the interests of good corporate governance and due diligence, the procurement department will examine company details for evidence of unsuitability such as bankruptcy, criminal conviction or failure to pay taxes and may examine annual accounts to check if a company is financially sound. It is equally important to check their technical capacity and track record in the market and the clinical engineering department may look for user references. Some organisations keep a list of approved suppliers, not to debar others but as an administrative aid. However, it is important to check all details are current. Equipment lifetime would be reduced if a manufacturer were to cease trading unexpectedly as it often becomes difficult to obtain consumables and equipment support. The requirement for safeguards will vary globally, depending on the standards of governance at national and local level.

A preliminary listing of the relative prices of shortlisted tenders supports the elimination of any clear high-cost outliers. However, a financial evaluation should not be aimed solely at identifying the lowest cost bid but at identifying best value for money in the provision of function, operating costs, maintenance and support. The contract award will be decided by arriving at an overall score using a weighting scheme that includes lowest price alongside other criteria for determining which is the most economically advantageous tender (MEAT) to the purchaser. Bottom-line prices are often presented differently by each supplier and need to be interpreted to accurately compare costs. For example, a specific physiological monitoring function may be included in the basic price by one manufacturer but be separately quoted as an add-on by another. In some systems, available functions are included as part of the installed software but are disabled unless paid for. In others, the extra facility may be a plug-in module. Sometimes the actual facilities and configuration options available only become clear to the purchaser during clinical or technical evaluation. The issue can become even more complicated when trying to work out lifetime costs, where options on both maintenance and consumables can further confuse how to compare different models. So although information in the tender may appear superficially to be purely financial, detailed technical knowledge is often needed to obtain a complete picture. A completed PPQ provides further background on device compliance with standards and regulations and also on more basic issues such as safety and product upkeep, including the provision of maintenance support, spares, the lifetime over which the product will be supported, software support issues, final disposal (see Chapter 12) and support for clinical training (see Chapter 7).

### 6.4.2 Clinical Evaluation

The aim of clinical evaluation is to establish how easy end users find it to use the equipment in practice and how well it meets their clinical needs. Issues to examine include the following:

- *Clinical functionality*: Although requirements should have been set out in the device specification, there is a level of detail where particular needs may not have been captured. For example, a specification might identify the space into which an item of equipment must fit but not mention that the room also needs to be large enough for occasional procedures with anaesthetic equipment. Putting equipment into the clinical environment where possible provides a good test of its clinical utility and of any practical difficulties in its use.

- *Ease of operation and use*: Including ergonomic factors such as control layout and other interface factors. Evaluation of these factors is inevitably subjective but users must be able to justify why they have attributed a particular score. Users should be careful not to be biased by recent experience with existing equipment or previous experience with older models and should be encouraged to take a fresh look at different manufacturer's products.

- *Suitability of different consumables and their relative costs*: Some devices use dedicated consumables but others generic items that cost a lot less. Part of the clinical evaluation is to see whether there is any difference between consumable items from different sources such as probes, and technical assistance may be required from the evaluation work stream to confirm compatability.

Clinical evaluation should be approached systematically. A set of users representative of application areas across the organisation should be chosen. End users need basic training in how to use devices being evaluated, so manufacturer support for this phase is vital and application specialists will assist users in getting the best out of equipment on evaluation. A checklist should be drawn up for every device evaluation, for consistency of approach. Clinical users may also spot potential hazards and design aspects which could prove problematic and it is important to record and act on these. If a potential problem is spotted, then it may take some time for a manufacturer to remedy it. It is not a good idea to rely on a verbal promise of a remedy before ordering the equipment, unless there is no practical alternative.

### 6.4.3 Technical Evaluation

Technical evaluation is the purchasing organisation's main opportunity to examine the potential purchase in depth, to check whether it really will deliver what the organisation is looking for or has any shortcomings or tangible advantages. The evaluation should also look at lifetime ownership, including the cost and future availability of spare parts. A technical evaluation will include at least the following:

- *Detailed comparison of the supplier's equipment specifications* set out in their tender document against those set out in the organisation's purchasing specification. This should include both essential and desirable elements. The aim is to identify and score how well the device meets user specifications and to explore any queries, shortcomings or other concerns with the supplier.
- *Test and inspection by qualified engineers* and technologists, preferably individuals familiar with the technology. This may include visits to sites where the equipment is in use or to demonstrations at a manufacturer's premises and will incorporate performance tests. Pre-planning is crucial, so that all necessary testing can be completed in the time available and specialist measurement equipment, simulators or phantoms can be made available. Generic evaluation checklists may already be available for some types of equipment, as might independent performance data.

- *Examination of technical literature and manuals,* to make sure they are clear and provide the necessary information.
- *Review of the costs and operational implications of maintenance,* with an estimate of the cost of replacement parts and accessories and of any specialist support equipment required.

Practical tips to help with this process include the following:

- *Reuse evaluation checklists where possible.* As topics such as electrical safety and other health and safety concerns are common to all devices, organisations will find it helpful to establish a library of checklists.
- *Establish at the outset if the item being put forward for evaluation is the same as the item that will finally be supplied* and, if not, how it will differ. A manufacturer may be planning to upgrade a device shortly and the organisation should seek to avoid ending up with some devices built to the old specification and others which are new, unless all can be upgraded to the same construction or performance specification. For example, one hospital purchased 20 operating theatre lights of nominally the same type, to find that there were several different and incompatible circuit board designs. This made dealing with breakdowns more difficult and costly and increased theatre downtime.
- *Establish compatibility with other equipmen*t in the surrounding environment, particularly where devices are sensitive to vibration, radiation or electromagnetic interference, and check for any other needs such as noise emission restrictions for MRI installations.
- *Maintenance evaluation should look at* ease of cleaning and disassembly for user replacement of parts; the quality of technical manuals and user instructions, together with the availability of maintenance and test software; and ability to reload or upgrade software after system failure or hardware change. Engineers should look for any design features that might lead to early damage or failure, including any particularly relevant to local conditions or intended use.

The likely costs of device maintenance may become clearer once the organisation understands how the device is constructed and what the replacement cost of individual sub-modules will be. Although a full risk and cost benefit analysis of maintenance options may not be carried out until after equipment is ordered, comparative costs for various manufacturers' support and maintenance regimes can be estimated against potential options for third-party outsourcing and in-house maintenance (see Chapter 9). Such an analysis may have been done previously for similar equipment and can provide a good basis for estimation. The ability to support and maintain equipment may be influenced by factors including the geographical location of a supplier's manufacturing, organisational and research base, the proximity of a service agent to the point of use and whether work is carried out by a directly employed workforce or through third-party agents. The purchaser may also have had previous experience with the company or their agent, and establishing a supplier's track record with other customers helps to indicate which maintenance options should be pursued.

### 6.4.4 Post-Tender Negotiations

During the evaluation phase, there may be discussions with suppliers in order to clarify or supplement the contents of their tenders, but in the European open and restricted procedures, the European Commission has issued a statement on post tender negotiations in which it specifically rules out any negotiation on price [6]. Its aim is to prevent further discussions on price or any other aspect of the tender that might distort competition or involve discrimination. In practice, unless a tender specification is straightforward, suppliers may come up with different approaches a purchaser would like to explore. For example, an imaging system supplier may include in their response details of an innovative image analysis package that offers the purchasing organisation unforeseen opportunities to improve the service. Other suppliers may be working on similar approaches, but the purchaser cannot determine this without running the risk of breaching the strict rules of the tender process. Suppliers may also offer additional goods or services at this stage in an attempt to secure a contract. There can therefore be a tension between running a fair process and achieving the best outcome. One way to resolve this dilemma is to make a judgement as to whether the original tender specification remains adequate and whether the outcome continues to be likely to give good value for money. The purchasing organisation then has the choice of either continuing with the original tender or of halting the procurement process and restarting it with a different specification. This is a complex area and specialist procurement advice is essential when considering how to deal with the unexpected.

### 6.4.5 Award

Once all information has been obtained, tenders are scored against financial factors such as purchase and lifetime costs and on compliance with the specification, taking into account the results of technical and clinical evaluations. Supplier reputation and stability and compatibility with an organisation's existing equipment can be brought in if referred to in the specification. A weighted scoring system provides a transparent and auditable process in case of challenge. If a scoring system is too rigid, however, it may generate perverse results. For example, one device may score very high on some counts and very low on a smaller number of others, resulting in an average score that beats a device which has an acceptable score over all evaluation criteria. Ratings cannot be based purely on numerical estimates and overall judgement is also important, so there should be some flexibility in interpreting the outcome. However, the further any decision deviates from that suggested by the numerical score, the more likely it is to attract the attention of auditors or bring the scoring process itself into question.

When a decision has been made, the contract will be offered to the successful supplier and unsuccessful suppliers will be notified. At this stage, it is particularly important that the decision-making process has been open and transparent and has followed legal requirements and the organisation's policies and procedures and that appropriate records cover all key parts of the process. Under European law, a 10 day standstill period is required before a purchase can finally be made, in order to give time for an unsuccessful supplier to submit an appeal over a breach of tender process. The remedies directive [7,8] allows suppliers or contractors to take action against individual purchasers in the high court, which has power to award damages if a contract is already awarded or to suspend the tender process if it has not. In specific circumstances, the European Court of Justice can overturn a contract.

An organisation need not act on the outcome of a tender. Its financial circumstances or clinical needs can change unexpectedly, though this wastes a lot of time and may be costly to a supplier. An extreme example would be where a company had tendered to equip a complete hospital but the scheme was stopped at a late stage. Under some national jurisdictions, the supplier may be able to seek redress, underlining the need to adhere to clear processes and keep comprehensive records.

It is possible to consider separately at this stage the possibility of leasing or other finance options, such as whether or not maintenance is included. Procurement of leasing finance will be by a second tender, with its own specification and complex set of possible outcomes. The chosen equipment supplier may provide this finance if it can make the best offer but equipment sales and financing functions are separated and evaluated separately to avoid the accusation of pricing bias by suppliers who know their equipment will be leased. Purchasers going down this route are strongly recommended to obtain independent advice, as exploring and evaluating the range of leasing options requires specialist knowledge of both finance and equipment markets.

## 6.5 Collaborative Procurement

Two or more hospitals can work collaboratively to form a group or consortium to purchase devices and negotiate contracts. A consortium should be able to achieve economies of scale in any tender process and thereby gains a negotiating advantage. One model is a regional body whose stakeholders are also shareholders and which aims to release resources for reinvestment in improved patient care and provide a better procurement service to staff.

A consortium may also have wider social aims than just economy and value for money, such as seeking to ensure its suppliers operate to certain employment and environmental standards. It can also provide better personal development and career opportunities for its staff. A consortium may operate through one or more lead hospitals or become an independent organisation. Similar large-scale procurement functions can be provided by private sector organisations, which may extend to independent equipping of hospitals or provision of equipment management services. The ultimate development of the consortium concept is national procurement, as occurs in some countries where all nationally funded hospitals are equipped by central allocation.

For consortium or large-scale purchasing to be effective, participating organisations need to be able to obtain clinical commitment to the outcome of tendering exercises. Local freedom of choice can be accommodated to some extent by developing framework agreements to cover similar devices from different manufacturers, for example, where a government decides to encourage investment in an expensive healthcare technology such as PET imaging or radiotherapy treatment. Purchasing can also be supported by independent evaluation centres or organisations. These review the specification and performance of different makes and models of equipment and can also produce template specifications. Organisations such as the National Institute for Health and Care Excellence (NICE) in the United Kingdom and not-for-profit organisations such as the ECRI Institute (see Chapter 13) provide information that save effort locally, nationally and internationally.

## 6.6 Summary

In this chapter, we have seen how to assess medical devices for their suitability for use in healthcare organisations. We looked at choosing a device to meet routine equipping and replacement need and considered how new technologies can be approached. We showed that, even for routine replacement needs, there is a tension between staying with familiar models and tendering for different technologies in the wider market place. We introduced the advantages and disadvantages of tendering and described the specification, tendering, evaluation and purchase processes in some detail.

## References

1. European Council. Directive 2004/18/EC of the Council of 31 March 2004 on the coordination of procedures for the award of public works contracts, public supply contracts and public service contracts. *Official J.*, L134, 114, 2004.

2. UK Statutory Instruments 2006 No. 5. The public contracts regulations 2006. http://www.legislation.gov.uk/uksi/2006/5/pdfs/uksi_20060005_en.pdf (accessed September 06, 2013).
3. World Trade Organisation. The plurilateral agreement on government procurement. 2012. http://www.wto.org/english/tratop_e/gproc_e/gp_gpa_e.htm (accessed September 06, 2013).
4. The Royal College of Radiologists. 2005. BFCR (05)1. Standards for ultrasound equipment. http://www.rcr.ac.uk/docs/radiology/pdf/standardsforultrasoundequipmentjan2005.pdf (accessed September 06, 2013).
5. British Nuclear Medicine Society. 2004. Gamma camera, PET scanner and data processor system tender questionnaire. http://www.bnms.org.uk/resources/gamma-camera-tender-documents.html (accessed September 06, 2013).
6. European Council. Directive 93/37/EEC of 14 June 1993 concerning the coordination of procedures for the award of public works contracts. *Official J.*, L199, 54–83, 1993.
7. European Council. Directive 92/13/EEC of 25 February 1992 coordinating the laws, regulations and administrative provisions relating to the application of Community rules on the procurement procedures of entities operating in the water, energy, transport and telecommunications sectors. *Official J.*, L76, 14–20, 1992.
8. European Council. Directive 89/665/EEC of 21 December 1989 on the coordination of the laws, regulations and administrative provisions relating to the application of review procedures to the award of public supply and public works contracts. *Official J.*, L395, 33–35, 1989.

# 7

# Equipment Training for Clinical and Technical Users

## 7.1 Introduction: The Need for Training

Training prepares users to operate and get the most from equipment. It is an integral part of providing optimal diagnosis and therapy, and minimises safety error risks to patients and other staff. Training varies significantly in scope, from showing end users how to operate a few functions on simple devices to equipping technical support staff with the comprehensive technical and operational knowledge they need to work with and repair complex equipment. This range is illustrated in Figure 7.1. Adequate resources and a systematic approach are required to plan, deliver and record training locally, supported by a good knowledge of who needs training and how it can be delivered. Local training is underpinned by wider education and training and input from equipment manufacturers and suppliers.

Inadequate training is known to cause adverse events. These vary from a minor inconvenience to serious injury and death, and cost money to deal with (see Chapter 3). Because effective training significantly reduces overall risk, it is no surprise that regulators and insurers insist on it. For example, UK regulator, the Care Quality Commission [1], expects workers to be trained to deliver clinical care safely and effectively, as does the NHS Litigation Authority (NHSLA) which runs a pooled insurance scheme with financial incentives for organisations to set up and improve systems for staff training and competence [2].

In this chapter, we consider the training needed by patients/carers, clinical staff and technical professionals operating and maintaining medical equipment. We look at how to deliver it and at who should train. We look at the organisational problem of delivering user and technical training to multiple users and discuss how training can be prioritised, delivered and recorded. Finally, we consider how training can be assessed and its value determined.

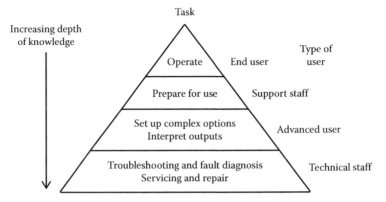

**FIGURE 7.1**
Different types of medical equipment users and their areas of competence.

## 7.2 Who to Train and What to Learn

There are three general types of medical device training. *End user training* is for those who operate equipment as part of a clinical procedure. They may also set it up, check it or carry out routine user maintenance. This group includes a growing number of patients and carers operating devices such as nebulisers and pain relief devices at home but mainly covers clinical staff such as nurses, doctors, therapists and healthcare scientists using equipment with patients. Such training must provide enough background knowledge to support safe working. *Train the trainer* courses teach clinical and technical staff how to train others in the use of a specific device. *Technical training* is delivered to technologists, scientists and engineers responsible for installation, testing, maintenance, quality assurance and user training and often covers details of clinical operation.

### 7.2.1 Clinical End User Training

Clinical end users need to know how to operate medical equipment for its intended clinical purpose, whether taking an image or cutting into tissue. This group includes doctors, nurses, radiologists, radiographers, sonographers, anaesthetists, physiotherapists, clinical physiologists and many other healthcare professionals. These professionals should have received some general training on the principles of operation, mode of application and major risks of the medical devices which they use, as part of their general training, and need to build specific skills and competences on top of this core knowledge. Generic training in electrical safety, the safe use of medical devices and relevant policies can be given at initial staff induction and refresher courses.

This element of generic training is often delivered by a member of the clinical engineering team. Refresher training might also be given to specific groups where staff turnover and device numbers are high, for example, infusion devices for nurses and foetal monitors for maternity staff.

All permanent and temporary staff using medical equipment for clinical care must receive training on a specific device before they use it. Training should be planned before new devices are introduced and when a staff member joins the team or changes responsibilities. Figure 7.2 shows how this might be done, with achievement of competency recorded in an individual personal profile listing all the equipment a user needs to be trained on. Equipment users, both temporary and permanent staff, then have a professional duty to keep their skills up to date. Individuals should have their training needs assessed annually and receive any necessary updates as part of their continuing professional and personal development.

End users must be able to use an item of equipment in normal operation. They should also be able to cope with malfunctions and handle incidents

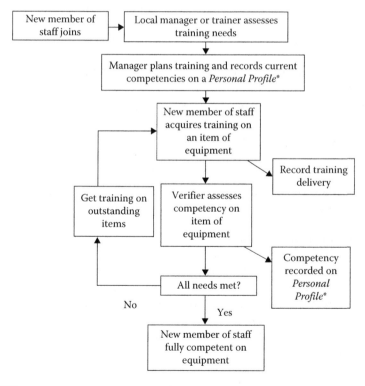

**FIGURE 7.2**
Assessment of training needs for a new member of staff. (*Personal profile is an individual's training record.)

safely, undertake decontamination and carry out any other duties required under the organisation's equipment management systems. In summary, they should be able to

- Distinguish differences between models, where these affect the safety of device function
- Connect the device to a patient effectively and safely
- Fit and use relevant accessories
- Read indicators and set controls appropriately
- Where relevant, show a patient or client how to use a device

If a device malfunctions, the user should be able to

- Recognise and correct malfunctions (or withdraw the device from service)
- Put contingency arrangements into operation
- Report malfunctions to the correct responsible person

If an incident occurs, the user should handle it correctly and

- Understand the importance of reporting device-related adverse incidents
- Be familiar with the organisation's reporting procedure

To meet infection control requirements, a user must

- Know when and how to decontaminate the device
- Be able to decontaminate the device following manufacturer instructions

Users should also be familiar with wider equipment management issues and their own roles and responsibilities.

Individual training needs depend on a person's prior experience, role and the devices they use. Capturing training needs is a local exercise, unless an organisation has strictly defined roles and standardised equipment. It is performed ideally at annual review and whenever a new make or model of device is introduced. Roles that need education and/or training to support them include the following:

- Responsibility for making the clinical decision to use equipment
- Deciding on equipment configuration and control settings
- Setting equipment up and operating it

- Interpreting readings, results or outcomes
- Devising protocols for equipment use and providing support to users

Local roles can be defined and used to set training requirements. For example, identifying the tasks and treatments a nurse in a high dependency unit delivers then indicates what equipment she/he will use and the level of training someone at this grade requires. Local clinical supervisors can work systematically through all the roles staff members take to map individuals to equipment types. This then shows what equipment training a role requires. A healthcare assistant may set up equipment prior to operation and perform basic checks, a middle grade nurse may operate straightforward functions of the device, and a senior nurse may configure the way it works and solve problems with its use. Once roles and requirements are clear, individual assessment will show what training remains to be done.

### 7.2.2 Training Patients and Carers

Increasing numbers of patients are supplied with electronic medical devices for use at home. These include continuous positive airway pressure (CPAP) units, nebulisers, transcutaneous electrical nerve stimulator (TENS) units, physiological monitors for heart rate or oesophageal pH and home dialysis units. Patients and their carers receiving these devices will require training by healthcare professionals or application specialists in their use, including how to carry out pre-use checks, identify problems and report breakdowns and incidents. The competence of patients and carers should be assessed, and they should sign to say that they have received training. Since the patient also has to undertake to return the device for maintenance and when they have finished using it, and verify that they understand that it remains the property of the healthcare organisation during the loan, training and sign-off form an important part of the equipment hand over procedure.

### 7.2.3 Specialised Technical Training

Knowledge of the general principles of operation, maintenance and testing of medical equipment forms part of the vocational foundation onto which training on specific device types can be built. Scientists or engineers working with major items of equipment, for example, in imaging or radiotherapy treatment, need extensive device-specific training in operation and quality assurance and will also benefit from knowing about the detailed technical operation of the equipment they work with. Technologists need detailed device-specific training on maintenance, testing and repair, but not all technologists need be trained in all devices. It is usually effective and efficient to train one or more in-house staff as specialists in specific device types whilst ensuring others can provide backup. A clinical engineering service might, for example, employ three people whose areas of expertise are ultrasound, ventilators and infusion devices, respectively, but each may be sufficiently familiar with the other two areas to solve day-to-day problems and provide competent cover for the expert.

Technical supervisors should ensure staff are adequately trained and qualified for the equipment they work on, identifying each person's training needs and arranging both general and specific training. Supervisors should also evaluate the quality and appropriateness of the training staff receive. There are no set guidelines for the frequency of technical training. However, a pragmatic minimum level of retraining is that it should be done at least when hardware or software is updated or when personnel or staffing levels change so that fewer people have knowledge of particular equipment. Technical support staff also have a duty to keep themselves updated, both as new technologies and devices are introduced and as new concepts and approaches to the management of particular device types and risk are developed. For example, new risks might be identified from the experience of others, including groups of equipment prone to failure or misuse, and exploring these issues with colleagues is an effective way for staff to expand their knowledge and identify good practice which they can copy and use to improve services. Training on new makes and models must be undertaken repeatedly during a technical career, and managers should encourage technical staff to attend professional update events and provide financial support or other incentives for staff to learn and apply new ideas to their work.

Manufacturers usually, but not always, provide some level of technical training for the servicing and support of their products, which the organisation is likely to have to pay for. This training may take place in the organisation, at another healthcare site, on company premises or at a separate venue arranged by the company. It costs hundreds or thousands of pounds per person, and the most cost-effective way to buy training is to include it in the specification for new equipment, alongside end user training, since the best opportunity for the organisation to negotiate a good price is at the time of purchase. A common aim of manufacturer technical training is to equip staff to carry out front-line servicing and routine maintenance, as part of a cooperative service contract whereby local technicians carry out routine tasks and only call on company engineers to solve problems that are not easily resolvable; see Chapter 9. The length of this training may vary from hours to weeks depending on equipment complexity and a trainee's prior knowledge. It is cost and time efficient to limit the number of people trained on manufacturers' courses because of their expense and problems covering front-line work if too many technologists attend courses simultaneously. Knowledge can be shared through on-the-job training within the organisation. Competence to act as a technical trainer under these circumstances should be assessed by the technical supervisor and will depend on individual experience, performance and the completion of professional development.

### 7.2.4 Underpinning Scientific and Technical Knowledge for Clinical Engineers

We have already mentioned the need to provide generic training to users. Such training is also needed by clinical engineering professionals at an early stage of their career and is part of the underpinning knowledge on which specialist training then builds competent and safe practice. This academic and vocational knowledge comes in two stages: (1) academic achievement at certificate, diploma or degree level, usually in electronic and electrical engineering; and (2) vocational education about medical devices in a clinical context.

Training of biomedical equipment technicians at *associate* level builds on secondary level science and a practical education in electronics (stage 1), with a vocational introduction to healthcare applications and supporting technical studies at foundation or basic degree level that takes the equivalent of two academic years (stage 2). For *practitioners*, the educational requirement is increased to full degree level (stage 1), with on-the-job modules and practical experience equipping trainees to work relatively unsupervised (stage 2). The entrance requirement for training as a *clinical engineer* is a good honours or master's degree in a relevant engineering subject (stage 1), followed by one or two year's vocational education at master's level alongside supervised training, and then a two- or three-year period of advanced training (stage 2). Accreditation of prior experience can also contribute to making an individual eligible for a course at higher level, together with a demonstration of their capacity for further academic study, as might be found, for example, when taking in a trainee who has already been through an industry or military training scheme.

For biomedical equipment engineers at practitioner level, about half of a typical vocational course will be spent on specialist biomedical equipment issues, with remaining elements split between basic engineering and science and general health-related topics. At associate level, there will be a greater stress on applied elements but with a similar mixture of engineering knowledge and applied healthcare skills. Clinical engineers include more on equipment and risk management, technology assessment and innovation. Vocational training schemes are discussed further in Chapter 13.

Short vocational courses help with stage 2 training and cover topics such as medical equipment safety testing and basic anatomy and physiology. Entrants can also gain background knowledge through technical or scientific conferences organised by special interest groups, and medical equipment companies may also sponsor annual meetings in clinical engineering which are suitable both for trainees and for established professionals updating their knowledge.

## 7.3 How to Train

The best approach to teaching and training is to use a combination of techniques matched to the needs of students, with frequent assessment of progress. People learn in a variety of ways:

- *Visual learners* prefer to see information. Some individuals respond more readily to diagrams and pictures, and others like to read text.
- *Auditory learners* prefer to hear information.
- *Kinaesthetic learners* prefer to learn by touching, manipulating and doing.

Common teaching methods with medical devices include the following:

- *Lectures* are good at reaching large numbers for education and updates, on topics such as general device operation and electrical safety. Attention can be difficult to sustain particularly where knowledge or ability varies, and lectures rarely engages kinaesthetic learners to any great degree.
- *Tutorials* and small *seminars* allow more individual interaction and can cope better with varied skills, abilities and learning styles. *One to one* sessions are often used to train healthcare professionals in advanced techniques.
- *Structured* and *self-managed learning* allows students to work at their own pace and can incorporate interactive and engaging material on technical topics. It is well suited to computer-based and online presentation and assessment.
- *Hands-on activities, simulation events* and *problem-based learning* are usually enjoyable and productive and are a good way to develop practical skills.
- *Videos and web applications* are expensive initially but reduce the ongoing cost of training. Updating material is a significant overhead.

Sources of teaching material include the following:

- *Manufacturer's training material,* often available to download in its entirety.
- *National schemes* may exist, such as the UK NHS Managed Learning Environment (MLE), allowing access to online equipment courses.
- *Professional body resources* covering the principles of device operation and reporting on user experiences and safety issues.
- *EMERALD and EMIT* European-funded projects [3,4] with online access to image databases and learning and reference materials for clinical engineers and medical physicists.

For lower-risk medical devices, users may be given ad hoc training *on the job* by a more experienced user, who is likely to be either another clinician or a member of the technical staff called out to assist. Such events should be reported to the user's manager or documented in a training record, to flag up any need for further training or competency assessment of the user. Ad hoc technical training is a simple and potentially effective route by which knowledge is shared when the opportunity arises, that is, when a problem needs to be solved. This process is effective in passing on knowledge about complex procedures and difficult situations. Trainers need to be aware that users instructed in this way may not assimilate all the required knowledge and should check final understanding. Training on higher-risk devices should only be given by accredited trainers, with careful documentation of what has been done and assessment of final user competence.

## 7.4 Organisation and Delivery

The organisation and delivery of training within a large organisation is complex and time consuming. It is usually a shared responsibility between end user departments, the clinical engineering service and organisation-wide training functions. Many healthcare organisations set up a medical equipment training section to provide core training on common items and help departments commission training from manufacturers and third parties. These services are often run by a mixture of clinical and technical staff, sometimes with access to dedicated training rooms or simulation facilities. In the United Kingdom, the National Association of Medical Educators and Trainers (NAMDET) is a specialist society supporting these roles. An organisational medical device trainer is likely to have a background as a technologist, nurse or engineer and will need support from other staff. Local trainers train staff or patients in specific areas of equipment or practice. For example, physiotherapists may train people to use devices at home, whilst practice development nurses and physiologists train fellow professionals.

Manufacturer and third-party training is less work for the organisation where courses run at regular intervals and training content is standardised. It can be expensive however, and there may be difficulty releasing staff for training particularly if delivery is limited to the normal working day or is only provided off site – a ward cannot be left without nurses, for example – entailing multiple presentations of a single course. This is a problem for all centralised training. A single one size fits all training course means that some staff may not have the background understanding required to attain competence and proficiency, underlining the need for subsequent local assessment of competence and the availability of top-up training.

Application specialists are employed by the manufacturer to provide end user training in the context of clinical procedures. They may also provide *train the trainer* courses to teach a cohort of experienced clinical staff how to train new users and will supply supporting material to help with this. Training can then be cascaded locally or through the organisation. Internal training by suitably educated trainers can offer flexibility in programme content and delivery hours, especially where courses are competency based. One possible pitfall is that any misconceptions on the part of the trainer are likely to be passed on to the users they train, so their capability and comprehension must be checked. In-house trainers are generally far more effective if they have general training in how to train, as this enables them to respond better to individual needs and circumstances. Similar considerations with cascaded training apply to the sharing of technical service training on equipment. There is a benefit to cascaded training, for the trainer, in that their involvement in teaching others consolidates their own knowledge by testing and stretching it in response to questions and difficulties experienced by the trainees.

When introducing new equipment across an organisation, such as a defibrillator, the timing of training has to be right. If it is too early, some staff may forget aspects of their training by the time equipment is introduced. If it is too late, the rollout will have to be delayed. Careful planning is needed also to coordinate the build up of training capacity with its delivery.

Where a lot of training needs to be done, a risk assessment helps to prioritise what is delivered, and to whom, within available resources. A pragmatic solution is to define levels of risk (extreme, high, medium and low) and assign each device model or class to one of these groups. Priority is then given to ensuring training is completed on all extreme and high-risk devices. Performance indicators for training should demand higher percentages of training compliance for higher-risk devices, typically 100% for extreme and high-risk devices. In practice, equipment users are likely to have experience with lower-risk devices already or can have their training brought up to date by local refresher training, leaving instruction on high and extreme-risk categories to be delivered by accredited trainers.

## 7.5 Training Records

Training records provide proof of an individual's competence and professional development for both audit and administrative purposes. They should include the nature and date of training and records of attendance signed by trainees or proof of satisfactory completion for online training. A record should also be kept of the outcome of any competence assessment made at the time of training and of final sign-off by a competence assessor or verifier that the individual can use the equipment safely and effectively in clinical practice. The trainer or organisation should maintain copies of training materials such as presentations, handouts and question papers for future reference, in case an individual disputes the training they received.

It is helpful if individuals have their own record, either paper or electronic, to show what training they have received and when they were assessed as competent. This record also acts as an individual training plan and, if reviewed at annual appraisal, can help to ensure that an individual's competence is kept up to date. Checklists are a good way to record competence against completed tasks or demonstrations of knowledge.

Local training records provide assurance that a satisfactory scheme addressing training risks is in place. However, fragmented local records are difficult to audit for the organisation as a whole, and it might consider capturing all training and verification sessions in one database. This function is included in the scope of some medical device management databases or, alternatively, might be provided as part of a global organisational training database setup to record details of all individual clinical, technical, professional and managerial training. Such institutional records of training may be held and administered by a combination of the organisation's human resources, clinical engineering, risk management and the clinical service departments. Locally, the form in which an individual's personal record is kept is decided either by organisational policy or by discussion with their line manager. For technical training, supervisors must keep training logs and assess competency, so that individuals can be authorised formally to work on particular items of equipment. These will include training certificates and proofs of competency from manufacturer courses.

## 7.6 Assessment of Training and Its Effectiveness

Competency assessors or verifiers are trained staff, typically approved through the medical device trainer, clinical engineering or another approved training lead, who check whether users understand what they are doing and are competent to operate a device or carry out a defined procedure. They must be experienced staff who are familiar with local procedures. Assessment tools are developed either by the manufacturer or in-house and can cover various competency levels on the same equipment.

Assessment helps to pick up inadequately trained users and, by extension, any problems with a trainer or the content of what they are delivering. The quality of training can be measured in the short term by testing whether individuals who have gone through it met the required learning goals. However, the desired long-term outcome, in which all users operate equipment effectively, is more difficult to measure. The organisation needs some way of assuring itself that users develop and retain their competence. An indirect method of monitoring performance is to look at reported equipment faults for evidence of user errors and at incidents attributed to inadequate training. In practice, training on existing equipment is often only updated when a serious adverse incident occurs, whereas best practice is to act early on information from near misses and minor events to seek to prevent major incidents happening at all (see Chapter 3).

Assessing whether individuals are competent on a particular task is a wider question than whether they have completed training successfully. Competence in the use of lower-risk items, or after a regular update, may be assumed by virtue of training having been given or following a direct assessment. However, clinical users operate in a pressured environment and deal with unpredictable situations, so ability is best assessed after a period of supervised practice through review by a qualified trainer or clinical supervisor. Various levels of assessment may be gone through to reach full competence in complex tasks.

After training has been carried out, it is possible to assess how much has been learnt by using tests or direct questions. Computer-based training provides an ideal opportunity to build in assessment via multiple choice questions. Practical exercises provide a means for assessing progress and identifying any problems a trainee may have in understanding what to do. These can involve a trainee dealing with simulated faults planted in a device by a trainer, for example, by use of incorrect accessories. For simple procedures, a demonstration of use to a supervisor may be sufficient for them to sign off the trainee as competent. In complex tasks, testing may give an indication of how successfully knowledge has been imparted but will not assess a trainee's competence at the task itself. To demonstrate this, the trainee will have to perform a number of procedures under increasingly lower levels of supervision, with a greater number of procedures being undertaken and signed off at each stage before they are assessed as competent. The most complex scheme is where the trainee initially watches the trainer perform a number of procedures, assisting to a greater degree each time, before then leading on parts of the procedures whilst being supervised or observed by the trainer. Finally, the trainee performs each task unsupervised, with the trainer checking the outcome. When sufficient procedures have been completed successfully, the trainee can be deemed to be competent. What constitutes a sufficient number will depend on the complexity of the task and the variety of problems likely to be encountered. At one end of the spectrum, for example, trained interventional cardiologists may be required to perform 50 or 100 cardiac catheterisations before they are accepted as competent in that procedure as the consequences of failure are so severe. If the scope of competence is chosen correctly, two to five repetitions are sufficient for most complex procedures, but there may be several to master for any particular device. Competence may also be assumed to lapse if the procedure is not repeated regularly or an update is not taken on time.

The effectiveness of technical training could theoretically be investigated by monitoring the number of equipment incidents and breakdowns before and after training. This is easier to do for end user training, where the number of trained users can be in the hundreds or thousands and where a consistent cause of breakdown or incident, such as repeated failure due to training shortfalls, can be isolated and monitored. However, it is difficult to draw statistical conclusions for technical training, given the small number of staff and pieces of equipment that fail. Overall effectiveness is best monitored by assessing the competence of each individual after training, by using tests or problem solving as outlined earlier, and by occasional inspection of repaired or maintained equipment by an independent competent person before it is returned to service. An early warning sign of lack of individual technical competence is repeated repair jobs on the same item in a short period of time, and detailed analysis can determine whether this is due to poor training or a lack of application by the individual. At an organisational level, benchmarking groups provide a comparison with other services (see Chapter 14), and suitable measures and peer review can be used to check whether outcome measures such as the percentage of equipment breakdowns in a year accurately reflect service effectiveness.

## 7.7 Summary

In this chapter, we discussed why training is needed and who needs it. We considered clinical end users, patients and carers and technical specialists. We introduced a number of ways to deliver training, whether given by in-house or external staff, from lectures to individual electronic learning. All this requires dedicated staff and resources to plan and deliver. We explained the importance of having training records for operational, audit and governance reasons, and of monitoring the effectiveness of training through competence assessment.

## References

1. Care Quality Commission. The National Standards. http://www.cqc.org.uk/public/government-standards (accessed on September 06, 2013).
2. The NHS Litigation Authority. Home page. http://www.nhsla.com/Pages/Home.aspx (accessed on September 06, 2013).
3. EMERALD 2. EMERALD e-Training resources in medical physics, free online version. http://www.emerald2.eu/Emerald2/ (accessed on September 06, 2013).
4. Tabakov, S. Special issue: Learning in medical engineering and physics (Editorial). *J. Med. Eng. Phys.*, 27(N7), 243–247, 2005.

# 8

## Assessing Maintenance and Support Needs

### 8.1 Introduction

Effort put into caring for equipment keeps it working properly and reduces unexpected failures and breakdowns. It also improves patient safety and continuity of clinical care. Clinical end users and managers often assume that the best way to look after medical equipment is to take out a manufacturer's comprehensive service contract and renew it promptly each year. In many cases, this is certainly effective and can provide peace of mind. It is however not always the cheapest, most appropriate or even the most effective way to support equipment and reduce downtime. Active management of equipment maintenance leads to substantial economies and also improves service quality, by decreasing equipment downtime and reducing the risks associated with equipment failure or malfunction. This is achieved by mixing three elements – manufacturer servicing, independent or third-party maintenance provision and support from suitably equipped in-house engineering services – in an evidence-based approach.

End users and managers understandably focus more on what equipment is being used for than on the details of its maintenance and how it can be optimised. Clinical engineers have the knowledge and experience to help set up, deliver and monitor a maintenance and support regime that gets the most from limited resources. They can provide expert input into maintenance contract negotiations, advising users and managers from a technical perspective and analysing the cost-effectiveness of proposals and ongoing service delivery. To put this into perspective, manufacturers typically charge around 8%–10% of the purchase cost of an item for a comprehensive annual service contract. For a large hospital with assets worth tens of millions of pounds, this can add up to an annual cost of several million. Experience shows that savings and efficiencies of 1%–2% of the purchase price are possible without affecting clinical services, by negotiating down contract prices or varying the level of cover. Using in-house or third-party providers can reduce costs by a further 2%–3%, especially where evidence is used to match the level of support to the risks of failure and can simultaneously reduce equipment downtime.

Maintenance is not the only support that equipment needs. By itself, it cannot guarantee trouble-free equipment operation – no contract can prevent damage due to misuse, for example. An overall support system should provide all the assistance users need, including training in the use and care of their equipment, equipment libraries, help desks, service visits and any other means of ensuring equipment is regularly inspected and correctly operated. Support can range from mundane but essential work to ensure that facilities for charging portable equipment batteries are adequate, through to proposing and managing the replacement of unreliable equipment. These activities require additional funding and resources but can more than pay for themselves by reducing maintenance costs and avoiding disruption of clinical services.

The risks of inadequate maintenance fall into three categories: clinical, health and safety and corporate, and were introduced in Chapter 3. In this chapter, we aim to stimulate practical thinking about how an organisation uses its resources to support medical equipment in a more cost- and risk-effective way. We consider how an effective support regime can be set up, looking at how to decide which elements of maintenance and support are best delivered by an in-house team, the original equipment manufacturer or a third party.

## 8.2 Balancing Elements of the Maintenance and Support Process

One aim of effective maintenance and support is to maximise the time equipment is available for clinical use and ensure continuity of clinical services. Another is to maintain the best performance possible throughout the equipment's lifetime, whilst minimising clinical and safety risks and the costs of keeping equipment in service. Achieving both aims requires constant juggling of activities, time and money across the different processes involved. An effective maintenance and support system has several different elements:

1. *Upkeep of equipment* is the responsibility of end users, who should check equipment frequently for correct function and any damage and also perform any regular user maintenance tasks specified by the manufacturer. Part, or all, of this role may be delegated to technical support staff or clinical engineers.

2. *User support* involves the provision of training, advice and assistance to the equipment user and may include the management of spares and consumables.

3. *Quality assurance tests* are carried out routinely by expert users to verify that equipment is performing to specification; an example would be the work done by physicists to check image quality on CT scanners.

4. *Preventive maintenance* aims to keep equipment working well by periodic observation, testing and proactive care. Actions arising from a preventative maintenance visit might include resetting any drift in calibration, routine replacement of parts subject to wear and tear and correcting any damage or deterioration unnoticed or unreported by users. At a more technical level, electrical safety testing involves the use of special test gear to verify that an item is electrically safe to use. It often forms part of routine in-house maintenance in some hospitals, as does electrical safety inspection which looks for equipment damage that can threaten electrical safety.

5. *Breakdown maintenance* is repair of a device after it has partly or completely failed, followed by its subsequent return to service. Many reported breakdowns are actually due to user inability to operate the device correctly or failure of an associated accessory or consumable. Troubleshooting is the first step of assessing whether a breakdown is spurious or a genuine device failure. It can take skilled staff some time to diagnose and repair a genuine fault, making it difficult to continue delivering the clinical service. Service continuity arrangements in the event of breakdown are therefore essential and may be far more important than the ability to repair equipment quickly. Contingency planning helps the clinical service prepare in advance to cope with equipment breakdowns and may involve spare equipment and accessories, the ability to borrow equipment at short notice or the adoption of alternative clinical procedures.

One way to guide resource allocation across these different support areas, for any particular type or model of device, is to look at risks, costs and benefits. The principal questions which such an analysis needs to resolve in the area of maintenance are as follows:

- What type of maintenance is required?
- What should it involve? Are inspection, testing and routine replacement of parts recommended and supported by evidence? If so, to what extent should they be undertaken?
- How often is maintenance to be done?
- Who is to do it? Manufacturer, in-house service, third-party providers?

If preventive maintenance is carried out too often, it will waste skilled time, money and materials that would be better employed elsewhere. If it is carried out too aggressively, for example, by periodically dismantling inherently reliable equipment, faults can be introduced which become risks in themselves. Electrical safety testing (see Appendix B) is another area which can consume considerable time for little benefit. Conversely, too little maintenance is likely to result in loss of clinical service due to equipment breakdown or even lead to clinical incidents, resulting in claims for negligence. High levels of equipment failure due to inadequate preventive maintenance and support waste time and money, with disproportionate time spent firefighting and dealing with emergencies. Matching maintenance to equipment is an ongoing effort which, in the airline industry, for example, has greatly improved reliability and reduced costs [1].

## 8.3 What Options Are Available for Preventive Maintenance and Support?

All equipment requires some level of safety and performance checking. At one extreme, the simplest low-risk equipment requires little more than pre-use checks by end users, coupled with relatively infrequent safety checks by suitably qualified personnel, with no regular invasive maintenance necessary. This is commonly known as *fix on fail* maintenance, although the relatively low cost of simple items such as nebulisers may simply mean that they are withdrawn from service on breakdown rather than being fixed. Alternatively, items that are low cost, operationally important and numerous, such as gas regulators, may be given an in-service lifetime by making their *in-date* status clearly identifiable by labelling or colour coding, after which they are withdrawn from service. Simple items are not the only ones assessed as *fix on fail*. Static equipment that is regularly used by clinical staff may need no regular technical maintenance other than periodic electrical safety checks. Fixed patient monitors are an example of this type of equipment as there is a strong probability any problems will come to the attention of users during normal operation. The level of care such units need often depends on the quality of user training and their inherent reliability rather than a predetermined maintenance interval.

At the other extreme lies equipment which needs regular planned preventative maintenance and careful attention, such as anaesthetic machines. These are life-support items, have moving parts and seals that wear out, and are used in an invasive patient environment (the operating theatre) where reliability and electrical safety are paramount. Such devices require regular replacement of parts at relatively frequent service intervals, together with rigorous and regular mechanical and electrical safety tests and performance testing.

In between these extremes lies a broad swath of equipment where service specifications from end users and expert advice from experienced engineers help identify how best to support equipment. An assessment tool and procedure is helpful in ensuring consistency of approach, and a written record provides the basis for evaluating outcomes and for future learning, as well as making the process transparent to auditors by recording the reasons why a particular judgement was made. Failure to maintain equipment adequately may be cited as evidence of negligence in legal claims: If a decision on how to maintain an item is logical and made by a transparent and structured process that would stand up to peer review, the organisation is likely to provide a good defence.

## 8.4  Assessing Maintenance and Support Requirements for Particular Devices

In most countries, healthcare organisations have a legal duty to keep medical equipment working safely and fit for purpose. One implication is that every item on the medical equipment inventory must be assessed to identify, for each group of devices, the level and type of maintenance and other support required, who will provide it and how often. The scope of this exercise needs to be wide as incidents are caused too by problems with unglamorous items of equipment such as wall mountings and trolleys that may be considered more as accessories than as anyone's maintenance responsibility. A standard form can assist with this task and facilitate a more objective assessment, and an example of such a form, and the rationale behind it, is given in the succeeding text. Although this process is deeply rooted in risk assessment, it is not a rating exercise but one where the form acts as a checklist to make sure all relevant factors affecting a decision are taken into account. Also, unlike a numerical risk assessment, the process does not produce a ranked outcome. A single answer to some questions may define the type of maintenance necessary. The main factors defining the risk posed by the failure of a particular item of equipment include the following:

- *Direct clinical consequence of equipment failure*: Such as a treatment withheld, an operation cancelled, radiation exposure to a patient or incorrect diagnosis of a disease.
- *Clinical service impact*: The effect on the clinical service of procedures being curtailed, constrained or unavailable.
- *Potential health and safety risks*: Possibility and consequence of injury due to mechanical failure, electrocution, etc.
- *Service interruption*: What is the clinically acceptable downtime for the equipment compared to the likely downtime in the event of a failure? Is substitute equipment at hand? Are alternative diagnostic procedures or treatments available?
- *Is repair of the equipment essential* to the service? Is service provision totally dependent on the function of a single device?

- *Can substitute equipment be brought in*? Are devices available elsewhere that can be used instead? Can the clinical work be rescheduled to another facility? Are spare devices or sub-assemblies kept elsewhere in the organisation for, by example, clinical engineering? Can the service provider or manufacturer find equipment at short notice? Substitution may be difficult to arrange for highly specialised equipment that is not widely available.

- *How can downtime be minimised*? Can manufacturers' response times be supplemented by in-house support, or vice versa, in the case of breakdown, by developing cooperative contracts, by the use of help lines or by having external engineers working on-site?

- *Legal considerations*: Are there any specific requirements covering this type of equipment? For example, ionising radiation legislation or electricity regulations may make specific inspection, maintenance and testing regimes obligatory.

- *Physical nature of the device*: Portable equipment is susceptible to physical damage, loss or theft, particularly if based on computers or used outside the hospital. Are special contingency arrangements needed?

- *Recommended maintenance*: Does the manufacturer recommend a particular regime for good reasons? is evidence available?

- *Age of the equipment*: There is likely to be an increasing probability of mechanical or sensitive component failure with age.

- *Equipment history*: The service history provides an evidence base for the level of past failures and will show whether the equipment has been maintained appropriately.

- *Level of user understanding*: Where equipment is not used correctly, or user understanding is poor, it may need more frequent attention. Where it is used by expert technical, scientific or clinical groups, the need for routine maintenance can be reduced by engaging users to carry out regular inspections.

- *Resources available in the organisation*: Are spares and test software available on-site? Is technical expertise available to troubleshoot and carry out minor repairs?

- *Special considerations*: Do any other issues affect this particular type of device?

Table 8.1 presents a form which can be developed to suit each organisation. Its final column highlights issues to consider against each factor. When completing an assessment, it is important for audit purposes to show every factor has been considered, by including a comment against each point. One issue – the existence of a legal or formal requirement to maintain equipment – may seem to make any further evaluation unnecessary, but, even in this event, general support is still required and also legislation or guidance may not

**TABLE 8.1**

Template to Help Plan the Type of Maintenance and Support for a Medical Device

| Factors to Be Considered | Comment | Notes: How Factors Might Affect the Decision |
|---|---|---|
| 1. Clinical *patient risks* (direct consequences of equipment failure or non-availability on patients or treatment) | | If equipment failure might impact significantly on patient treatments, then investing effort in pre-use checks or tests, routine inspections and maintenance by skilled personnel at more frequent intervals could be worth doing to reduce risk to an acceptable level. If failure is critical, for example, in life-support environments, the risk may only be acceptable if spare equipment or parts are readily available on-site. |
| 2. Clinical *service risks* (effect on the service of losing access to equipment) | | If equipment failure would impact significantly on a clinical service, by, for example, extending patient waiting times beyond accepted targets, then the same steps as for factor (1) should be considered. There is also a risk to the reputation of the service and possible long-term negative consequences. |
| 3. Clinically acceptable *downtime* for the equipment<br><br>Likely downtime in the event of a failure for:<br>• Equipment<br>• Clinical service | | Short downtimes are inevitable, but this factor needs to consider how long the service is sustainable without the equipment. Where an item is likely to be out of use for some time and the workload can be managed using other equipment and extended working hours, issues such as overtime costs and staff and patient inconvenience are likely to limit how long the situation can continue. It may also take some time to set up and then to dismantle alternative arrangements, so planning arrangements in advance can reduce the impact of any unforeseen failures. |
| 4. *Alternative capacity*<br><br>Spare equipment available?<br><br>Alternative diagnostic procedure/substitute treatment available?<br><br>Loan equipment available? | | If adequate replacement equipment or alternative procedures are available, downtime may not affect clinical risk significantly. This is a straightforward way to cover high-risk services. Costs can be reduced by ensuring that separate departments with similar services share common backup equipment, and calculations of the probability of simultaneous failure based on historical breakdown records can help to estimate how many backup devices are needed to reduce risks to an acceptable level. |

(*continued*)

**TABLE 8.1 (continued)**

Template to Help Plan the Type of Maintenance and Support for a Medical Device

| Factors to Be Considered | Comment | Notes: How Factors Might Affect the Decision |
|---|---|---|
| 5. Any specific *legal, regulatory or recommended requirements* for the type of equipment? | | Legislation or regulation may require measurements, such as outputs of ionising or non-ionising radiation. Regulations and standards do not tend to specify intervals for activities. Discussions with professional colleagues are necessary to establish the range of actual behaviour as opposed to any stated ideal. Organisations will need to justify any variation from accepted practice. |
| 6. Any *special considerations* not covered earlier? | | Hazard notices from a manufacturer or notified body may imply specific maintenance procedures are required. Also, equipment that has been involved in clinical incidents may require special maintenance regimes, to reduce the risk of a similar failure occurring. Some types of monitoring equipment may require regular testing/replacement, for example, oxygen fuel cells. |
| 7. *Components likely to fail*, for example, moving parts/ seals/rechargeable batteries | | Manufacturers may recommend regular replacement of items. The organisation needs to take note of and generally follow these suggestions but may choose to shorten or extend replacement intervals based on a careful analysis of removed parts or experience of failures in service. Some of the areas where this approach is worth adopting include the following: |
| Is the device associated with mountings, wheels, pendants or trolleys? | | All *mechanical moving parts* will wear and so carry a risk of failure, as do electromechanical parts such as motors. |
| Portability of device? | | The performance of *rechargeable batteries* will degrade after a number of cycles, and fluctuations in temperature and humidity affect the reliability of electronic components such as capacitors and integrated circuits. Routine testing of battery capacity and/or prophylactic replacement might be needed. Regular charging regimes will maintain battery life. The chemistry and properties of *materials* change over time, so elastic seals will become less flexible and be liable to leak. *Pendants and trolleys* are awkward to check and often get overlooked yet cause a significant number of incidents. |

**TABLE 8.1 (continued)**

Template to Help Plan the Type of Maintenance and Support for a Medical Device

| Factors to Be Considered | Comment | Notes: How Factors Might Affect the Decision |
|---|---|---|
| | | Physical damage will be more likely if the equipment is *portable*. Regular inspections and the prompt replacement of cases and external parts will limit possible further damage. It is more difficult to exclude the possibility of internal damage and consequent hazards due to dropping or rough handling. Loss is more likely, particularly if devices are used outside the hospital. Portable devices may resemble computers and thus attract thieves. Enhanced security arrangements, location tracking and provision of spare units might be considered to reduce risks. |
| 8. *Service history* <br> • Age of device <br> • Breakdown history <br> • Maintenance history | | A relatively new device should be more reliable and so could be covered by an ad hoc maintenance agreement. However, a history of frequent breakdowns may indicate replacement as a long-term solution, with high maintenance needs in the short term. Careful analysis of reasons for failure can reveal weaknesses in equipment design, which can be tackled by proactive replacement of failing parts and discussions with manufacturers about potential modifications. |
| 9. *Recommended maintenance* Intervals and extent of work required | | Manufacturers' recommendations must be taken into account, as discussed in (7) the previous text, but can be varied especially if a good evidence base can be built up for an alternative regime supported by a risk assessment. |
| 10. *Spares and software availability* | | Where parts and software are difficult to obtain, particularly for older or superseded items, more frequent maintenance or inspection will be needed to identify the need for replacements in good time. Buying in a stock of critical components before a supplier runs out can prolong the useful life of old equipment. For large and expensive items, organisations that have the resources to source parts from the original subcontractor or make necessary items directly will be able to continue using them well beyond any nominal life. |

(*continued*)

**TABLE 8.1 (continued)**

Template to Help Plan the Type of Maintenance and Support for a Medical Device

| Factors to Be Considered | Comment | Notes: How Factors Might Affect the Decision |
|---|---|---|
| 11. *Expertise and resources available* within the organisation and other maintenance suppliers | | Expertise is critical when considering how to set up support arrangements. In-house support is viable for more common devices but proportionately more expensive for specialist equipment. Manufacturer and third-party maintenance are viable at all levels and may be the only cost-effective solutions when a critical mass of equipment cannot |
| The range of possible providers of maintenance support | | be established. An exception to this is where particular individuals have the required expertise in-house, due to either recruiting staff from manufacturers or entering into cooperative maintenance agreements and training with manufacturers or other suppliers, when cooperative arrangements can deliver faster response and repair times and allow preventative maintenance to be done at convenient times. |
| Cooperative service contracts may be a good idea, either with external maintenance suppliers or through linking to another in-house organisation with a different range of available skills | | A high level of technical expertise relevant to equipment type or model may reduce the need for routine maintenance and allow maintenance regimes to be varied, particularly regarding the proactive replacement of parts liable to fail. The speed of response and fault diagnosis will increase as staff gain experience, but this can take years in any one organisation. |
| 12. *Support regime decision* Routine and breakdown maintenance? | | Decide on all forms of support, articulating clearly whether or not each one is required and, if so, at what level. |
| Internal maintenance/part replacement? Electrical safety checks/ inspections? | | Specify maintenance intervals and/or occasions on which each form of support is required and who is responsible. |
| User checks and maintenance? Quality control? User training? | | The amount of support users require and the level of input they can give to caring for equipment depend on their background training and level of specialist knowledge. |

specify how often or to what extent maintenance should be carried out. Apart from this category, there are no automatic decisions and all factors need to be taken into consideration when making a judgement about what to do.

## 8.5 Assigning Responsibility for Maintenance and Support

### 8.5.1 Introduction

Having decided on an appropriate maintenance and support regime, the next step is to decide who performs each aspect of the work. All parties involved in supplying information or making the decision – end user, budget holder, clinical engineering department and external provider – have a particular viewpoint. End users tend to support renewing with their existing provider, if they are comfortable with the support they are getting. Manufacturers and third-party providers look to make money from equipment servicing. So like the maintenance and support decision, allocation of responsibilities is best made on the basis of a risk and cost–benefit analysis which is made as objective as possible.

The initial choice for many organisations is between *insourcing* and *outsourcing* – whether to perform maintenance on-site using staff employed by the organisation or whether to commission servicing from the original manufacturer, their service agent or a third party. In practice, the decision is usually more complex than this, as various scenarios are possible whereby different providers cooperate to deliver equipment maintenance and support. One scenario is to have no service contract and to use in-house staff for preventive maintenance, front-line troubleshooting and minor repairs whilst calling on the manufacturer for ad hoc repairs. More complex possibilities include a manufacturer basing their staff in the organisation or running a cooperative contract where in-house staff screen faults before calling out the manufacturer who works with them to reduce downtime. Similar considerations apply to an organisation's relationship with any third-party maintenance provider. A different type of support is offered by some third-party organisations, who offer to commission servicing for all or part of an organisation's equipment base. This can be in specific areas such as intensive care or may cover a specific device type, such as anaesthetic machines. It may also be linked to a *managed service* equipment procurement and replacement scheme.

Further decisions about the division of responsibility between end users and technical support services for wider equipment support include who is responsible for everyday user-level maintenance, such as routine cleaning and regular checking of blood gas analysers, where clinical users may want to delegate this work to a technical service. Finally, it is important to be clear who decides whether technical support is provided in-house or by external providers. This decision is often left to a local budget holder who should take advice from clinical users and technical experts. Where an organisation lacks technical expertise, users should consider consulting external sources of advice, as quotations or other information from manufacturers or third-party service providers on technical requirements reflect their particular perspective and need to be scrutinised against the organisation's needs, using the concepts outlined in this chapter. This is particularly important if maintenance is extensive or costly, or the organisation is considering employing an agent to commission maintenance services on its behalf.

Some organisations delegate all or part of the equipment maintenance budget to their in-house technical service, which then makes the decision on levels of support in collaboration with clinical users. This approach can be effective in reducing overall costs, as a view can be taken across the organisation to provide equivalent levels of support to the same equipment in different departments, and individual service contracts can be combined to achieve savings based on an increased volume of work. Some specialist third-party organisations provide cover for specific device groups, such as imaging equipment, across multiple manufacturers at reduced rates.

### 8.5.2 In-House Maintenance Provision

Responsibilities for maintenance must take into account the aforementioned factors such as available in-house expertise, number of items in service and complexity and risk associated with the equipment. These topics are explored further in the succeeding text and in Chapter 9 for external contracts.

In-house breakdown maintenance nearly always has the advantage of proximity and flexibility, which together with local knowledge and contacts gives staff the ability to respond quickly to call-outs and perform front-line troubleshooting well before any external organisation can get a person on-site. Finger trouble and unidentified faults can be diagnosed quickly and consumables and accessories changed as necessary. Local technical staff get to know their equipment and how best to prevent problems, particularly where trained and supported by the manufacturer in a partnership arrangement. Internal technical staff can also provide ad hoc user training to correct problems with misuse and, by virtue of being on-site and available to support a wide range of equipment, pick up problems with unrelated items or advise clinical staff on equipment purchase and other issues. This is particularly true when supporting and advising on more common types of equipment. Where preventive maintenance is carried out by an in-house service, similar advantages of proximity and flexibility apply as when performing front-line support. Unplanned downtime can be reduced by using in-house resources to identify imminent failure, via proactive inspection and the purchase and use of dedicated test equipment.

For more specialised and complex equipment, the detailed technical knowledge available to in-house services will usually be inferior to that of the manufacturer or a specialist third-party provider, who both have the advantage of seeing a given type of equipment perform in a variety of settings and who can access extra resources such as specialised test equipment. Service staff from the manufacturer should be able to diagnose and deal with complex faults more quickly than in-house staff, once they are on-site, and be able to replace parts and assemblies more quickly. The quality of support staff is crucial, and experience with a manufacturer or third-party maintenance organisation will show whether their staff are as knowledgeable and capable as they should be. This is where it is helpful to have an in-house clinical engineering service which can work with end users to scrutinise the timeliness and quality of work done by external suppliers. It is not unheard of for an external service organisation to provide poor service or fail to resolve a complex fault on an item on comprehensive cover. End users need to monitor breakdown frequency and service staff attendance and be robust in challenging repeated failures.

A manufacturer will usually need to be involved when there is a serious fault involving breakdown of a major component or a fault that resists obvious troubleshooting procedures. An in-house service will need to have the facility to call them in if necessary. Manufacturers are increasingly building diagnostic functions into their equipment and seeking to offer remote services to predict when key components need to be changed and to quickly identify the cause of faults. Organisations will need to look carefully at what these services offer, and their cost, before committing to using them. Users face potential problems with data and network security if remote access is planned, and the information technology (IT) department will need to review and approve arrangements for access.

Availability and cost of spare parts is also a consideration. A manufacturer's service organisation is likely to carry a wider range of spares in stock than an in-house service, due primarily to cost considerations. Indeed, the price a manufacturer charges to in-house services for spare parts can be high, although costs can be limited by agreeing a pricing approach with a supplier prior to equipment purchase. Availability of spares can be enhanced by making a prepayment against which parts can be ordered with minimal bureaucracy. Sometimes identical components can be sourced directly but more cheaply from the same supplier used by the manufacturer or as generic components, as can accessories and materials. If parts are used which are not supplied by the manufacturer, the organisation will need to make sure parts are traceable and are equivalent to what is being replaced. A clinical engineering service can provide valuable support in this regard, and money can also be saved in the short term by replacing components or sub-assemblies rather than whole assemblies, where this is possible and trained in-house labour is available and cost-effective.

In summary, in-house maintenance support has the advantage of greater speed for front-line troubleshooting and can reduce costs for straightforward maintenance, whilst the manufacturer has the advantage of technical expertise for more complex faults and the servicing of specialist equipment. However, there can be financial and operational advantages from performing more complex servicing and repair in-house. How far the in-house service

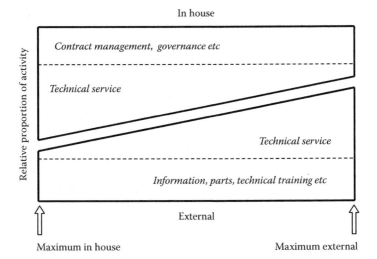

**FIGURE 8.1**

General spectrum of possible maintenance arrangements. The proportion of technical support between in-house and external suppliers can vary but neither the organisation nor the manufacturer can avoid some level of responsibility for maintenance activities.

is able or wishes to venture into the latter area, and for what equipment, depends on many factors. In practice, maintenance and support arrangements always involve some level of cooperation between the organisation and external providers, and Figure 8.1 illustrates this.

### 8.5.3 External Maintenance Providers

Manufacturers have their own servicing operations run on local, national or regional lines with mobile staff and specialised facilities. This section looks briefly at other options for external maintenance provision that might be met in practice.

#### 8.5.3.1 On-Site Services

Where a large amount of equipment from a particular manufacturer is installed on-site, having a member of their staff permanently seconded to the healthcare organisation combines the proximity of an in-house service with ready access to the skills and knowledge of the manufacturer. User support and training, and routine maintenance and repair, may be provided as part of a single package. Another advantage of having a manufacturer or other third party on-site is that responsibility for managing their personnel lies with the provider, which leaves the organisation free to concentrate on defining the service and monitoring the contract. However, suitable contract management skills need to be available.

### *8.5.3.2 Independent or Third-Party Maintenance Providers*

A healthcare organisation may ask a third party – one independent of both itself and the manufacturer – to maintain all or part of its equipment. Many such providers were started by one or more service personnel trained by a manufacturer who then set up an independent service competing initially on cost and customer relationship. An area where this approach is viable is ultrasound equipment servicing, where most machines operate on standard principles and offer broadly similar functions. As the size of an independent group grows, it can take on more makes and models and start to offer value-added services such as the provision of loan and exchange equipment. Using such a provider allows the organisation to amalgamate service support across a range of makes and models and lower overheads to a point where costs are reduced below those a single manufacturer can economically provide. Quality of service can also be high as the group accumulates experience across different organisations but it needs to be kept under review, as direct competition with manufacturer service organisations may lead to potentially restrictive practices by the latter such as making it difficult to access key spares or coming up with technological developments that are difficult to service without expensive facilities. A manufacturer may go into a strategic alliance with a successful group, subcontracting elements of its servicing work but retaining the management of customer contracts. Some very large operations have developed in the United States and elsewhere and are making significant inroads into the maintenance contract market, even to the point of remanufacturing parts to original manufacturer specifications.

### *8.5.3.3 Maintenance Contract Management Companies*

A second group of independent providers, set up by individuals with contract management skills acquired in a finance, procurement or facilities management context, offers to manage all or part of an organisation's equipment portfolio, usually by linking together procurement, maintenance and replacement activities. The link with equipment procurement, and the ability to tie in financing arrangements such as leasing, is attractive to organisations which have limited funds. Managed equipment services can also be restricted to maintenance only. Some companies will take on and manage existing contracts whilst making a saving typically of 10%–15%. This is achieved usually by tighter control of contracts, to ensure maintenance is not taken out on devices the organisation no longer owns, and by replacing comprehensive maintenance cover for all or some items by support from a manufacturer or third-party service organisation as and when required. In practice, such organisations make money from and are most interested in large items of equipment but will apply their skills to managing the numerical bulk of smaller items for a suitable reward. The healthcare organisation has to manage any contract with an external maintenance commissioning organisation very carefully to ensure that savings are real and are shared equitably over time.

### 8.5.3.4 Diversified Service Organisations

In the past, manufacturers concentrated on repairing their own products, but many are now able to adopt the role of multiservice providers and maintain whole ranges of equipment they do not themselves manufacture. Some equipment manufacturers who have become involved in private finance or other equipping initiatives have set up such initiatives by diversifying the equipment portfolio covered by their existing service organisation to take over management of a particular service function, type of equipment or management area. Examples include total management of a contract for infusion devices or intensive care equipment, particularly where a manufacturer has a substantial equipment base in a local service and can extend this to support other manufacturers' devices. They approach this practically by either recruiting staff directly, taking over departments with trained staff or acting as a commissioner to manage service contracts with other manufacturers. This hybrid approach combines the same advantages and disadvantages as for manufacturer servicing and commissioning organisation support discussed previously.

In summary, the advantages of third-party maintenance include diversity of choice and of equipment covered, with independent groups using risk management and diversity to reduce costs in the same way as in-house services. The effectiveness of working relationships between third parties and manufacturers is important and can only be assessed in practice. Looking carefully at the experiences of others who have been using a third party is an important part of deciding whether or not to use their services. As service organisations diversify into multiservice vendors, they take on a wider range of equipment belonging to more than one manufacturer, and the range and quality of these services need to be considered carefully before signing any contract to commission them.

### 8.5.4 Factors Affecting What Maintenance Is Undertaken In-House

The ability to take on equipment maintenance in-house depends on the type of equipment, the nature of the in-house service and the resources available to it and support from end users. If the support required for a particular type of equipment is appropriately provided by the in-house service, the next step is to assess whether existing resources such as labour, space and equipment are available and if staff have the necessary expertise to support the proposal. If not, the service will have to acquire it within the required time scale and financial constraints, by recruiting specialist staff or sending existing staff away for training. It is also important to find out if the organisational culture supports taking on the new work and if a persuasive case would be supported by senior management. Each of these elements is now considered in more detail.

### 8.5.4.1 Type of Device

It is not possible to set a clear boundary between what can and cannot be maintained in-house. Although the most complex devices are generally not maintained in-house, there are some significant exceptions, usually where individuals with specific expertise can be recruited. Devices such as patient monitors or syringe drivers are present in large numbers in acute care settings and are often looked after in-house. An alternative deal is offered by some manufacturers who provide a complete maintenance, repair and substitution service linked to purchase of a minimum quantity of consumables. In this instance, availability of finance is usually the deciding factor rather than equipment complexity, fleet size or servicing costs. Devices of reasonable technical complexity, present in moderate numbers, include items such as defibrillators, anaesthetic machines and ventilators. Although many in-house services cover this type of equipment, supported by manufacturer servicing training and by backup equipment, other groups are not confident taking on the clinical risks and technical complexities posed by these devices. At the top end of difficulty are x-ray equipment and radiotherapy linear accelerators where high voltages and ionising radiation present considerable hazards and the clinical risk of malfunction is high. Full service support is provided by those departments that have the expertise and inclination, whilst many others provide some level of front-line support in partnership with manufacturers. Other types of equipment are never maintained in-house because specialist expertise is essential and minimising downtime is a critical factor, including complex imaging and laboratory devices. In summary, the equipment most commonly selected for in-house repair is simple to maintain, easy to understand and straightforward to operate. It is present in sufficient numbers for clinical services to be covered easily by substituting a similar device in the event of breakdown and for sub-assemblies and parts to be swapped readily between devices as part of the fault-finding process. Multiple items bring economies of scale, reducing the time and expense taken to train and equip staff and the cost of mistakes made whilst gaining experience. Conversely, a poor candidate for in-house maintenance would be an expensive and complex piece of clinically vital equipment, with only one on-site, requiring complex skills acquired by extensive training and long experience.

### 8.5.4.2 Restrictions on What Can Be Done

There are some absolute restrictions on what can be done in-house. Special training and authorisation are required in various countries to work with medical gas, radiation and environmental cooling equipment. Appropriate professional status, qualifications and supporting equipment are required.

Even some relatively simple equipment such as weighing scales can be subject to regulatory control for medical use, making calibration other than by a registered test house unlawful. Another problematic area is mechanical equipment, such as patient hoists or equipment mountings, where consequences of poor maintenance can be fatal or lead to serious injury. For all these areas, an organisation will either have to set up a service in-house which conforms to the regulations or commission an external service who offsets the cost of regulatory compliance across a much larger customer base. There is a strategic choice for the organisation as to whether it wants to become a leading provider in one or more of these areas. It has to balance the reduced costs, potential external income and quality benefits from establishing an in-house service against the ability to offset at least some risk to an external provider. There are also technical considerations. The need for specialised test equipment and software is growing as equipment becomes more sophisticated, both for complete devices and for sub-assemblies. Software-driven tests are now very common, but without approved access to the manufacturer's servicing codes, an in-house service will be unable to perform them. Nearly all new devices are software driven and must be kept updated. On many systems, backup software is essential for repair and even routine maintenance. The in-house service will have to purchase or have access to this software and any associated dongles or test cards, and this needs to be agreed as part of initial equipment purchase.

### 8.5.4.3  Pressures to Maintain Equipment In-House

Less commonly, there can be an absolute necessity to maintain equipment in-house. A highly specialised clinical service might use equipment developed or modified by staff within the organisation, and there is little chance an external body would be willing to take on its maintenance. A clinical service may depend on equipment which is unsupported due to its age or because the manufacturer has gone out of business, yet must continue to use it to perform an essential clinical function until a replacement is possible. Distance too is a factor. Where a manufacturer does not have a service agent in a country, and the clinical service relies on old, second-hand or donated equipment, there may be no choice but to repair equipment in-house.

### 8.5.4.4  Nature of the In-House Service

Relevant factors here are the attitude and ambitions of the service – whether or not it considers itself suitable for a particular type of work – and its existing and available resources. Every in-house service is a unique collection of individuals with specific past experience and a historical set of attitudes. Its existing portfolio will give an indication of the type of work it has been willing and able to attempt in the past and hence what is likely to be taken on in the future without major reorganisation. The ambition of a service is often related to its size, and small departments will find it difficult to take on significant amounts of new work without advance planning and recruitment. The size of an in-house group must reach a critical mass before it can routinely develop new maintenance services with a minimum risk, for example, by asking existing staff to work additional hours. If significant expansion is undertaken, there may be difficulties in recruiting and training suitably skilled staff, and costs associated with building up an infrastructure of equipment and spares can be a high proportion of the existing budget. Operations already successfully providing larger-scale services may be keener to submit a business case to take on new work, including responding to tenders from outside organisations. Risks that seem formidable to a small service may be accommodated comfortably in a larger one because of their ability to redistribute resources in the short term to cover unforeseen difficulties. Similarly, a larger existing infrastructure provides economies of scale and scope for expansion. Further consideration of in-house services can be found in Appendix A.

### 8.5.4.5 User Buy In: Internal Politics and Culture

Internal politics within an organisation cannot be ignored when developing the business case for a new service. For example, if in-house facilities and expertise are available, cost-effective and operationally acceptable, it would appear to make sense to maintain equipment in-house and not take out an external service contract. However, this may not be the perception of end users or managers who for various underlying reasons may lean towards manufacturer servicing. Clinical users might regard an in-house service as less reliable or want to retain what they see as a higher degree of control over their service contracts and payments and preserve an existing direct relationship with the manufacturer. On the other hand, others might assume in-house maintenance is automatically cheaper or better because they know the workforce to be dedicated and not seeking to make a profit. Likewise, assumptions about cost-efficiency in the private sector are unjustified until the conclusion of negotiations over contract prices and conditions, accompanied by robust monitoring of contract performance and close control back to the contract terms (see Chapter 9). Prejudices can only be countered by objective assessment based on evidence, together with a forum for airing different viewpoints and investment in developing a transparent service and good working relationships. Some perceptions are based on historical events and not current reality and can be very difficult to change once established. Relationship issues may be rooted in long-standing agreements with manufacturers about what is seen as special support, particularly in larger institutions where clinical staff compete to get access to the latest technology and external scrutiny of cost-effectiveness may be seen as unwelcome interference. It is therefore politically necessary to convince all stakeholders of the rationality of what may be perceived as a high-risk approach and to understand and address their concerns. This can often only be done by producing evidence that the in-house clinical engineering service can or will add value in the long term, by helping to solve technical and management challenges and improving the quality of service to clinical users.

### 8.5.4.6 *Willingness to Change Existing Maintenance and Support Regimes*

There may be good reasons why existing arrangements have evolved in a particular way. However, all services can benefit from periodic review. Attitudes to growing and updating service provision become self-fulfilling: small-scale services remain small because they do not seek opportunities for growth or diversity and large-scale services continue to seek ever-increasing challenges and continued savings. Opportunities for a fresh approach include mergers or joint initiatives between previously separate healthcare organisations, where a unified approach to equipment management is sought, major staff changes and the introduction of new technologies or major developments in clinical services. Externally imposed budget cuts or efficiency measures threaten to reduce the quality of clinical services and often lead to a re-evaluation of existing practices. One practical example is how to encourage groups in the organisation to forgo a service contract where none is indicated. It is essential for clinical engineering to work closely with all stakeholders, explaining clearly the rationale behind the decision and how savings in expenditure from a cancelled contract can be reinvested to increase the quality of maintenance overall. An important point to clarify is how any savings will be allocated. At the opposite extreme, a clinical engineering assessment may indicate manufacturer maintenance is the best option, but end users fear being ripped off by contracts and repair bills. They may make unrealistic assumptions about the savings that can be made by an in-house service. Many end users have little idea of the overheads in personnel, equipment, management and space that are necessary to provide an in-house service. Here, a tabulation of the costs and resource required will help to form the basis of a convincing business case.

### 8.5.5 Changing Maintenance Regimes

If a decision is made to take on a new maintenance activity, the in-house service faces the operational risks and associated costs of training staff and setting up the new service and also has to deal with corporate risks including potential increased liability from taking on maintenance risks formerly adopted by the manufacturer. Similarly, where a manufacturer or third-party provider is asked to take on the maintenance of equipment previously supported by the in-house service, they are likely to insist that the organisation pays for an initial inspection or service upfront, and for the cost of remedying any defects.

### 8.5.5.1 Service Risks

The safest approach to changing to in-house maintenance for a particular equipment type is to increase the complexity of operations undertaken in-house over time, as workforce skills increase with experience. As a first step for less expensive devices, spare units are bought or older items kept and substituted for equipment removed from service for repair. As familiarity with the construction and failure modes of the devices increases, more ambitious repairs can be performed such as the replacement of sub-assemblies. There is a trade-off between initial investment in spare equipment and staff learning time, requiring a long-term vision which can be frustrated by changes in funding or workload. For more expensive devices, taking on elements of front-line support builds up in-house expertise alongside formal manufacturer training, prior to taking more maintenance in-house. Arrangements must be made to guarantee the availability of spare parts and specialist backup in case of significant failures and difficulties in their resolution. In some cases, even for quite expensive equipment, it might make sense to limit repair effort, and where repair cost is a substantial proportion of device cost, money saved by not having a contract can be used to buy spare units. Replacing and scrapping items that fail may be cheaper than running a service contract and manufacturers may offer part exchange of faulty units for refurbished items.

### 8.5.5.2 Staffing Risks

Investment in permanent staff brings management responsibilities and overheads. An alternative for new work is to hire staff on agency or short-term contracts, with an option to employ them permanently once the new service has operated successfully for a reasonable time. Another approach is for existing skilled employees to undertake the new activity, building on their local knowledge and experience, whilst using temporary staff as backfill.

### 8.5.5.3 Regulatory Risks

Agencies such as the UK Health and Safety Executive require equipment to be suitably maintained and safe for operation and may prosecute flagrant breaches. Clinical regulators are concerned with patient safety and service effectiveness, so they may inspect to make sure equipment is maintained. Insurers may also require evidence of maintenance before providing liability cover. Any maintenance arrangements should satisfy these requirements.

### 8.5.5.4 Liability Concerns

In the United Kingdom, the Medicines and Healthcare products Regulatory Agency (MHRA) recommends carrying out maintenance according to manufacturers' intervals and procedures [2]. The principal perceived risk from changing maintenance procedures or using non–original equipment manufacturer (OEM) components is not that alternatives will necessarily be substandard or less reliable – although for non-OEM parts, some genuine incompatibilities may exist – but that if general guidance is not followed to the letter, a contract or warranty may be voided unintentionally or the organisation will become liable for damages if an incident goes to court. Similar risks arise with any change of maintenance frequency away from the manufacturer's recommended service interval with changes to the recommended test and inspection schedule, or the use of in-house trained personnel who have not received formal training from the manufacturer. In most cases where evidence-based in-house operational decisions are made not to conform exactly to a manufacturer's formal procedures, this does not mean recommended tests and component replacements are ignored. In practice, changes mainly involve altering maintenance intervals and finding alternative suppliers of spare parts and consumables that are cheaper or can be delivered more reliably and speedily. In some instances, an in-house service may increase servicing frequency in the light of their own experience, to prevent premature failure. Good risk management is essential to successfully counter any litigation that raises these points as deficiencies, including compliance with essential requirements such as legislation and keeping documentation of the decision-making process and day-to-day maintenance actions.

## 8.6 Final Review and Decision Making: Deciding Who Performs Maintenance

As discussed earlier, the decision as to who will maintain equipment may be obvious according to the nature of the equipment, capability of the in-house service and culture of the organisation. The most interesting cases arise when either in-house or external support is viable and a rational decision-making process is needed to provide guidance and transparency. People involved in the decision-making process can range from a single clinical engineer or end user deciding on the arrangements for a simple low-cost item to a project team deciding on how to maintain a complete imaging service or new healthcare facility.

Factors affecting the decision can be tabulated on a pro forma. Many are similar to those used when deciding on maintenance requirements. Although a single factor can determine the outcome, it should be possible to perform a combined risk–benefit and cost–benefit analysis in order to come to a decision. Costs of the in-house service need to be included in their entirety, including wages, space, equipment, tools and training when comparing with external contract costs. Table 8.2 provides a template to help identify these costs. Benefits are not only financial but also operational and clinical, including anything adding utility or value to a service. Putting a financial value on a general benefit is difficult and subjective, as is ascribing a monetary loss to

**TABLE 8.2**

Cost/Benefit Considerations for In-House Maintenance Services

| | Elements | Comments |
|---|---|---|
| *Service costs* | | |
| Personnel | Technical<br>Managerial<br>Administrative | Costs need to include all employment overheads. Include agency staff at full cost. Allocate costs in direct proportion to staff time taken up by the new service. |
| Accommodation | Workshop space<br>Office space<br>Storage space | Space requirements should be included in the business case. Insufficient space may prevent service expansion unless extra locations or ways of using space more efficiently are found. Building and other space adaptation costs should be included. |
| Test equipment | General<br>Device specific | Test equipment includes general electronic items and also equipment-specific tools and software such as special testing circuits and programmes, phantoms, chemical sensors and flow measurement test rigs. |
| Tools | General<br>Device specific | Some equipment needs custom-made tools to disassemble, test or reassemble. |
| General support | Transport costs<br>(if applicable)<br>Administrative consumables<br>Organisational<br>management overheads<br>Quality management<br>system | Organisations charge a general overhead rate which must be included in any cost–benefit analysis. Registration to an external quality management system will have a cost but helps an in-house service comply with its own procedures. |
| Training | General – clinical/scientific/<br>technical device training<br>Manufacturer's device-<br>specific training<br>Legislative/health and<br>safety<br>Manual skills | Device-specific training can be very expensive for large items and should be negotiated as part of original equipment purchase. Any training costs must be included in a business case. Training *on the job* should be carried out by authorised trainers with staff competence assessed. |
| Software tools | Service software<br>System software including<br>equipment database | Costs include a proportion of hardware and software support contracts, with provision for regular updates and upgrades. |
| Consumables | Components<br>Workshop materials | Storage and deployment processes need to ensure adequate traceability of components. |
| Maintenance and replacement | Upkeep and emergency<br>repairs to accommodation,<br>test equipment, tools and<br>consumables | This includes the cost of regular test equipment calibration. |

*(continued)*

**TABLE 8.2 (continued)**

Cost/Benefit Considerations for In-House Maintenance Services

|  | Elements | Comments |
|---|---|---|
| *Service costs* | | |
| Outside repairs and assistance | Manufacturers' costs for ad hoc call-outs, spares and consultants or temporary staff to assist with problems or new developments | Estimates of the costs of ad hoc services can be difficult to make and should be based on the best available experience within the organisation or elsewhere. |
| *Service benefits* | | |
| Financial | Contract savings<br>Generic part sourcing | Contract savings need to be calculated in the context of their effect on any other contracts, for example, where the organisation gets a discount for aggregating multiple contracts. Base any parts savings on firm quotations. |
| Operational | Response time<br>Access to replacement equipment<br>Other contingency arrangements | Changes to response time are difficult to cost without knowing the effect on the overall service. Looking for a user estimate of value will provide the best quantification for a cost–benefit analysis. Cost any contingency arrangements such as out of hours working to repair critical equipment. |
| Clinical | Downtime<br>Access to fully operational equipment | Careful planning and scheduling of preventative equipment maintenance can reduce unplanned and overall downtime significantly. Any extra costs should be balanced by an estimate of financial benefits. Quantify the benefits of regular inspection and equipment care by looking at past failures and the cost of contingency arrangements. |

risk, and differences between individual perspectives give rise to the political issues considered earlier in this chapter. Combining these factors with the risks also mentioned previously forms the basis for the comparison and evaluation template in Table 8.3.

If a decision on how best to maintain equipment is marginal, it may be made clearer by looking at how it fits with the organisation's long-term strategy. For example, the in-house service may aim to expand in the long term to the point where it has taken on all additional economic work or be seeking to downsize by outsourcing difficult to support items where this brings financial and operational benefits. There is also a trade-off between short and long objectives, especially where a service has recently taken on a new facility or batch of equipment. In this instance, an assessment of the

**TABLE 8.3**

Factors Affecting Whether Maintenance Is to Be Undertaken In-House or Contracted Out

| Factor | In-House Service | Original Equipment Manufacturer | Third-Party Maintenance Provider |
|---|---|---|---|
| *Clinical risk of equipment failure* | Low-risk equipment. In-house service confident of knowledge, procedures and contingency arrangements and ability to cover liability. | High risk or liability. In-house service not confident of defending against any litigation arising from inadequate maintenance. | Medium risk, proven track record of supporting clinical services. |
| *Uptime criticality* | Front-line support capability to respond quickly and effectively within required timescales. | Uptime critical and in-house expertise limited or nonexistent. | An ability to respond more quickly makes this option more attractive. |
| *Availability of substitute equipment* | Large fleet of similar items. Interchangeable units on-site, hence minimum downtime. Interchangeable sub-assemblies, substitute testing. | Small number of units only. However, a large fleet may be totally managed by manufacturer or agent with economies of scale. Lack of substitute units or sub-assemblies, hence risk of extended downtimes. | Suitable for areas that have well-developed expertise and backup support. Some providers may provide exchange or replacement equipment under prior arrangements. |
| *Equipment complexity* | Equipment either of a lower complexity or where manufacturer supports in-house servicing. | Complex equipment requiring a high degree of specialised knowledge not available in-house. | Various providers cover equipment of differing complexities. |
| *Expertise available* | In-house knowledge and expertise exists or is easy to obtain. | Engineers knowledgeable and well trained. | Proven ability to support type of equipment being considered. |
| *Spares/consumables* | Guaranteed pricing or discount available on parts. | Spare parts quickly available. Mark-up on parts not excessive. | Spare parts quickly obtainable at reasonable pricing. |

*(continued)*

**TABLE 8.3 (continued)**

Factors Affecting Whether Maintenance Is to Be Undertaken In-House or Contracted Out

| Factor | In-House Service | Original Equipment Manufacturer | Third-Party Maintenance Provider |
|---|---|---|---|
| *Existing relationships with supplier* | Good working relationship with manufacturer's/agent's technical services. | Near, good communications and delivery/travel routes for service team short. | Near, good communications and delivery/travel routes for service team short. |
| Proximity to manufacturer/agent | Access to software/service tools. Manufacturer/agent no longer exists or manufacturer service ineffective. Distant, agent has no local service team, must be shipped abroad for repair. | Reliable service operation and field engineers well supported and able to communicate effectively with clinical users and in-house contract manager. | Reliable service operation. Field engineers well supported and able to communicate effectively with clinical users and in-house contract manager. |

equipment inventory will highlight the most worthwhile areas for improvement. It is likely that an overall assessment of the organisation's maintenance and support arrangements will identify savings that can free up resources for investment in additional services, or be returned as overall savings.

The wider consequences of every short-term decision should also be considered, to see if it will have any longer-term consequences. Reactive change for short-term economic advantage may look beneficial but be difficult to reverse. A service agency may, for example, offer low prices to gain business, which then results in the dismantling of a good in-house team. Such a team will be difficult to rebuild if the decision is reversed in future years. Likewise, a manufacturer's service operation may suffer if customers make large-scale cuts in demand to save money. The organisation also has to be careful when cutting back in-house manpower in lean times as costs may simply be externalised onto more expensive manufacturer maintenance contracts that come out of other budget lines in the organisation, resulting in an overall increase in organisational spending.

## 8.7 Summary

Equipment maintenance is expensive. Careful consideration of what maintenance is actually needed can save money which can be spent elsewhere to reduce risks, for example, in clinical user training and support. The most cost- and risk-effective maintenance regime for each device should be determined using a transparent process by qualified staff. Maintenance and support of medical equipment can be provided in many different ways by different providers, principally the original manufacturers, in-house services and third-party maintenance and managed equipment providers. Each has particular advantages and disadvantages that are specific to the equipment, the manufacturer and the healthcare organisation. In general, manufacturers have the advantage when it comes to complex equipment that is vital clinically and present in small numbers. In-house services can have the advantage of proximity and local knowledge, provided they have the technical capacity to provide a service.

There are many factors to weigh up in making these decisions. The broad thrust of the advice we have given in this chapter is to

- Consider objectively the consequences of equipment downtime and what contingency measures can be taken to substitute an alternative item of equipment or procedure
- Look at the breakdown history of this type of equipment in general and the item in particular
- Look carefully at the costs and benefits of possible maintenance options in the light of regulation, standards and overall maintenance strategy
- Plan what maintenance will be done and who will provide the different elements
- Review maintenance experience and needs periodically to see whether other options would be more cost-effective

## References

1. Brinkley, W. The effects of maintenance on reliability. http://www.reliabilityweb.com/art08/effects_of_maintenance_on_reliability.htm (accessed on September 06, 2013).
2. MHRA. DB2006(05). *Managing Medical Devices*. Reviewed in 2013. Medicines and Healthcare Products Regulatory Agency, London, U.K., 2006.

# 9

## Maintenance Contract Management

### 9.1 Introduction

In theory, an organisation could commission a service agent to provide any pattern of equipment support they wanted. In practice, options are limited to what can be delivered affordably and sustainably. Most medical equipment is purchased for the long term, and service contracts become an arena for negotiations between the interests of the purchaser and those of the maintenance provider. For the user, factors determining which type of contract to take up are similar to those used when assessing maintenance and support needs as set out in Chapter 8, namely, the nature of the device, facilities available and the costs and risks of equipment failure. The organisation must balance clinical, health and safety and corporate risks against potential financial savings. For the service provider, any maintenance support must be delivered profitably in the context of their overall market and portfolio of contracts. The most expensive elements involved in delivering service contracts are people, parts and the logistics of bringing them together at the right time and place. Delivering standard packages of maintenance limits administration costs and provides greater predictability, helping servicing providers with service continuity. A good relationship between purchasers and providers is vital, with effective partnership working being in both their interests.

Larger healthcare organisations often run multiple service contracts related to particular models or makes of equipment alongside in-house maintenance and support. Service contracts are not only provided by equipment manufacturers, but also by third party suppliers, sometimes as part of a service covering a range of equipment types or locations or even as whole conglomerations of contracts across multiple organisations. Where large-scale outsourcing is being considered, this chapter should be read in conjunction with Chapter 4 which provides an overview of major outsourcing programmes or totally managed services.

We refer to the *clinical engineering* department throughout this chapter to describe two different but overlapping roles. The first is as a provider of medical equipment maintenance and support services to end users. This role supports a range of medical equipment within the organisation as an in-house maintenance provider. The term *clinical engineering* in this context covers all in-house support services, whatever department staff work in. The second role offers broader, organisational support with staff advising users, clinical managers and corporate functions on policy and practice in areas such as clinical governance, technology procurement and adherence to policy and standards. These roles overlap. As in-house service engineers develop knowledge on particular equipment, they generate data useful for reliability statistics and help to guide approaches to maintenance and replacement. External service engineers do not normally perform this monitoring function unless specifically commissioned to do so.

## 9.2 Maintenance Contract Management Life Cycle

### 9.2.1 Setting Up the Contract

Contracts, like equipment, have a life cycle that needs to be managed (see Figure 9.1). Initially, the need for a new contract has to be established and funding identified. Then the scope and nature of a contract has to be decided before reviewing available options from manufacturers, third parties and in-house services and then selecting which option to pursue. Prices, inclusions/exclusions and responsibilities must be negotiated once the

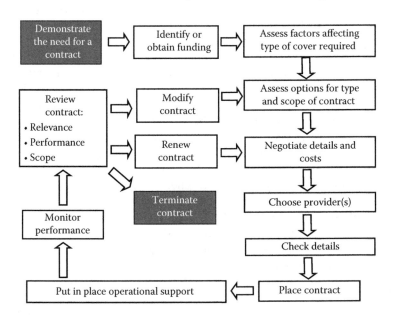

**FIGURE 9.1**
The contract management life cycle.

scope and provider of the contract are decided. After negotiations have been finalised, the contract must be checked to make sure the provider is legally liable for their repair work and penalties are defined for poor performance or premature termination. After a contract is placed, those in the organisation with responsibilities under the contract, including clinical users, must be told what they have to do and be given any training they need. Details of all servicing activity and equipment performance should be recorded and monitored. At least annually, contract arrangements must be reviewed to see whether they are still effective before renewal, modification or termination. We now consider this cycle in more detail.

### 9.2.1.1 Establishing the Need for a Contract

This topic is covered extensively in Chapter 8. If the need for a contract is agreed by the end users and clinical engineering service, after taking into account all the factors discussed in Chapter 8, then the process of looking for an appropriate provider and an appropriate level of contract can begin.

### 9.2.1.2 Identifying Funding

Resources for maintaining and supporting new items of equipment should ideally be identified before purchase, when new equipment funding is approved (see Chapter 5). If the decision-making group is concerned solely with capital purchases, it will not be responsible for allocating maintenance funds but may make the provision of a maintenance budget a condition of awarding capital funding. At the purchase stage, the manufacturer should be able to quote their rates for contract maintenance, but whole life support is likely to cost more and can only be estimated. Spending includes not only service contracts but also parts and labour charges for ad hoc repairs or accidental damage, payments for parts replaced during routine maintenance, training and other support costs. It would seem sensible for an organisation to automatically identify budget funding to cover the maintenance and support costs of new equipment as it comes into service. This is difficult to achieve over the long term, however, as budget setting is a political process which distributes finite resources, and nobody can be assured of guaranteed increases.

### 9.2.1.3 Allocating Funding

Responsibility for allocating a maintenance budget falls either to the user department or to a centrally managed budget. The user department must either allocate existing budgets or acquire new funding, whilst a central budget which may be held by the clinical engineering service, a group of services, the procurement department or a senior manager. How expenditure is controlled will contribute greatly to user perception and to the cost-effectiveness and efficiency of maintenance and support. The first question is how far to group or centralise maintenance budgets. Where a local service has control, they can pace expenditure to match variations in budget and demand, with procurement of back up spares reduced to a minimum when money is short. They can decide where to invest or reduce spending if equipment is more or less reliable than expected. With pooled budgets, the budget holder does not have such immediate feedback and it will be more difficult for them to appreciate service impact if equipment fails. Opportunities to use local expertise will be lost unless there is close working between users. Aggregating budgets over a group of services may create opportunities for risk sharing and economies of scale, but these are best realised where budgets are held in a central pool. The advantage of this arrangement is one-stop ease of use and the ability to assemble specialist expertise to monitor and manage contracts. However, close relationships with clinical users are needed to maximise benefits and ensure issues such as poor maintenance or equipment being withdrawn from use are addressed quickly and effectively. A contingency budget is helpful to cover the costs of unexpected repairs. A department can use this funding flexibly, for example, to replace an ageing low-value item rather than pay a significant amount for yet another repair. It can also move between in-house and external support for small-value contracts, as the demand for service and its own workload varies. Generally, access to a well-managed contingency fund for maintenance allows contract expenditure to be lower overall than it would be otherwise.

### 9.2.1.4 Deciding on the Provider

Possible providers of medical equipment maintenance and support include manufacturers or their agents, third-party providers or an in-house service. Manufacturers have the advantage of detailed knowledge of their products and access to expertise and specialised test equipment and spares and are able to provide a comprehensive and, in most cases, reliable service. However, third-party providers are becoming more common and are able to provide services based on types of equipment from several manufacturers or in specified clinical areas. On a larger scale, they can supply *contract bundles* where many or all of an organisation's contracts are managed by a single provider. These can provide significant savings over the approach of placing single manufacturer's contracts for each device type. There is further discussion of these types of service in Chapter 8. Deciding on the overall balance of provision is complex and iterative, closely related to the nature and scope of the contracts themselves. Where external service contracts are of sizeable cost, particularly in the case of multiservice vendors, tendering may be necessary. Discussions with potential providers gather information to help make the decision. This includes talking to in-house clinical engineering departments in both of their roles: first, to establish the level of maintenance, repair and support they can offer, and second, to see what expertise is available for contract monitoring and other specialist support. Information may be available in-house to guide what level of maintenance support to choose. Where maintenance is wholly or partially carried out in-house, it should be delivered to a defined level even in the absence of an explicit contract or service-level agreement. Hence, the considerations that apply to specifying monitoring and reviewing manufacturers' contracts equally apply to in-house services. As for an external contract, clear lines of operational and financial responsibility and a defined scope and frequency of the expected in-house service must be defined. Proactively managing these service aspects avoids uncertainty and delays in coping with breakdowns and other problems.

### 9.2.1.5  Reviewing Options for Contract Type and Scope

Having done the background work, the next step is to review the support levels and pricing structure of contract options. Although virtually any combination of service support can be negotiated, the types of contract available off the shelf generally fall into one of several categories. Typical examples are the following:

- *Platinum*: All breakdown maintenance, loan units, parts, accessories cover, annual or more frequent service visits, helpline, minor software upgrades
- *Gold*: As earlier but downgraded to reduce costs, for example, with major limited life items such as x-ray tubes or key parts excluded
- *Cooperative*: As for one of the two options earlier but with the initial response/assessment carried out by suitably trained in-house staff
- *Silver*: Service visits and a limited number of breakdowns or part replacements
- *Bronze*: Labour and parts for planned service visits only
- *Custom*: By negotiation, virtually any combination of the earlier and other options, including cover for accidental damage and pre-payment discounts on parts and call-out costs

### 9.2.1.6  Factors Affecting the Level of Cover Required

The minimal possible contract is usually routine maintenance or service visits only. These are predictable and so easier to cost, for both service provider and organisation, and to monitor for value for money. They also go a long way towards fulfilling many operational and risk management requirements, providing auditable proof of maintenance and reducing the risk of irreversible long-term equipment deterioration. Although regular visits should reduce the risk of failures in operation, they cannot prevent them entirely and the next level of contract specifies a given number of call-outs. Risks associated with any level of cover are mostly financial rather than operational, as individual call-outs and repairs beyond those covered may be very costly in the event of serious breakdown. On the other hand, this may never happen and significant money will be saved. Thus, the level of cover depends upon balancing probabilities of breakdown against the cost of cover. Where anything less than a full contract is proposed, the costs and conditions of emergency call-out and repairs must be estimated in advance based on prior experience, to compare likely and worst-case costs with potential savings. Manufacturers usually charge a number of elements to these call-outs: an hourly labour rate, travel time to and from the site, parts costs and sometimes an administration fee. There may be a minimum charge or set cost, which is why it is worth looking at different options to see which is the most economical in the long term. Clinical engineering should keep equipment history records for all equipment on contract, for both their own work and that done by external service providers, to provide information on the historical costs of in-house and ad hoc repairs. This evidence gives a basis from which to judge reliability and the likelihood of future failures and associated parts costs and, hence, choose potential options. End users will also have an idea of equipment reliability and be able to comment on operational support issues, particularly where they are significantly affected by equipment breakdown.

### 9.2.1.7 Cooperative Contracts

These place initial responsibility for dealing with breakdown call-outs in the hands of in-house staff, trained by the company. This can improve response time markedly and speedily resolves general faults which are due to finger trouble (inadequate operational knowledge of the equipment) or to failure of easily replaced minor accessories and consumables. Cost savings in cooperative contracts are skewed towards the contract provider. Customer discount is set at a minor proportion of the full contract price, typically 5%–15%, but the proportion of front-line work put in by customer staff compared to the service provider is likely to be somewhat larger than this. The customer may also have to bear the cost of in-house staff training by the company. These financial considerations can be more than compensated for however by a reduction in response time and speedier diagnosis of trivial faults resulting in less clinical disruption. Another advantage is that in-house staff can often identify serious faults that require company attendance, using manufacturer test routines and embedded equipment software, identifying early on any spare parts likely to be needed. In the long term, cooperation should help to keep contract provider prices down.

### 9.2.1.8 Breakdown Response Time and Hours of Support

Users and clinical managers need to identify how critical equipment is to the service and can refer to front-line evidence of the impact of breakdowns. A key judgement is whether money spent buying a particular guaranteed response time is well spent, in relation to the service downtime it is likely to save, as response time should match clinical need. If a clinic operates infrequently, there is little point in paying for an 8 h response if the service engineer cannot arrive reliably within a few hours and a guaranteed downtime of less than a week might be much cheaper. The times during which a supplier provides their service and what they consider as normal hours also have a bearing on how quickly things happen. When are engineers available to discuss problems and arrange call-outs? Are response times or downtimes reckoned in working days only, or do they include nights, weekends and holidays? The availability of two items is critical – knowledge and parts. Some large companies arrange telephone support from different time zones, which helps in-house engineers and their own service personnel progress repairs out of hours. Being able to order spare parts out of normal hours through a telephone hotline can reduce downtime when getting key components despatched and delivered earlier. However, a quick response does not necessarily involve a fast repair. Separate assurances are required for this. How long will a company attempt to solve a problem before escalating it to bring in their most experienced engineers? How good is their engineer backup? Will they provide substitute working equipment after a given time spent trying to carry out a repair and are there any associated costs or conditions for this service? If the process involves returning equipment to the company for repair, service or replacement, what are the typical and maximum turn round times? In summary, although it is not possible for a service support provider to be certain how long a particular repair will take, the purchaser should review their proposed arrangements to assess how effective these are likely to be and how well they match its needs. Other users' experience is likely to prove helpful in making such an assessment and asking for references from a new provider is always advisable.

### 9.2.1.9 Costs

Some costs could be overlooked when negotiating a contract, including the following:

- Breakdown call-outs beyond a specified number
- Labour for call-outs at unsocial hours such as evenings and weekends
- Travelling time for call-outs – this makes the proximity of the provider's service centre or engineer a factor in the decision
- Parts other than those routinely replaced, for which a requirement is identified at a routine maintenance visit
- Parts or labour for breakdown repairs above a specified cost
- Software or hardware upgrades
- Repeat attendances when equipment is not available for routine servicing, whether because of administrative errors at the customer's site or because it is in use to meet clinical need

### 9.2.1.10 Replacement of Expensive Parts

The life of durable items such as ultrasound probes, x-ray tubes and endoscopes depends on how heavily they are used and how well they are treated. Some maintenance contracts include a breakdown insurance option for expensive parts, where the purchaser pays a regular annual amount in anticipation of failure, but actual time to failure is unpredictable. Contract suppliers make a judgement of average lifetime based on multiple devices, a knowledge base against which a customer cannot compete. It may not be cost-effective to take out this kind of insurance unless the product is so new that no one is clear how long it will last, or the organisation has sufficient internal experience to judge that its own failure rate is lower than average. It is more likely to get better value by setting up its own internal insurance contingency fund, particularly if it has multiple units where replacement is in the order of tens of thousands of pounds and the likelihood of failure is high. Political support will be needed to do this and to overcome any operational obstacles allowing rapid approval of spending against this fund. Where this type of arrangement is in place, it is possible to take advantage of highly specialised third-party providers who carry out particular types of repair, on items such as ultrasound probes, more cheaply than the manufacturer who may only offer a service exchange. Users may need in-house support to manage flexible options to help take into account past equipment performance and consider warranty cover and other factors. Specialist companies and their stuff come and go, so obtaining background information on the quality of their work is also essential.

### 9.2.1.11 Agreeing Operational Responsibilities

Service contracts are only a part of the overall context in which devices are operated. Other responsibilities need to lie where appropriate for the provision of devices, maintenance and other support, including consumables and spares. For example, where a total management contract is negotiated with a supplier for a single device fleet – syringe drivers, for example – the cost of new or replacement units and maintenance is, by definition, largely borne by the company. Yet it may be operationally effective for some maintenance and support to be provided by in-house staff, to reduce response time, so even when much of the maintenance burden is adopted by an external provider, some contractual and operational issues remain to be sorted out. Relevant questions include how far are front-line staff and any in-house maintenance service responsible for fault identification and front-line troubleshooting? Will the manufacturer visit at intervals to service devices and repair broken equipment, or are devices despatched to them for repair? If the latter, who arranges this and bears the cost? Will items be despatched by in-house personnel or collected by the manufacturer? Also, as equipment manufacturing techniques change, the technical resources needed to carry out comprehensive maintenance and repair are becoming increasingly specialised, as discussed in Chapter 8. If front-line repairs and maintenance for a particular items are to be performed mainly or partly in-house, whether associated with a service contract or not, the resources to do this must be obtained from the manufacturer or their agent. In addition to appropriate service manuals and training, engineers need software codes to access engineering test software and system recovery software and install upgrades. It is becoming increasingly common to be unable to perform any meaningful service operation without these. Rights of access should be formally agreed with the company in writing as part of the service package or purchase agreement.

### 9.2.1.12 Check Details before Signing

It is important to make sure that all equipment serial numbers are listed on the contract and that these match those units actually present. This is particularly relevant for items with expensive interchangeable sub-assemblies, including monitors, ultrasound scanners and associated probes or light sources and endoscopes. External engineers or internal staff may exchange items without recording details in organisational or company records, and clinical staff sometimes move probes from one machine to another. Some companies only send out service engineers to a breakdown call after checking an item's serial number is covered by a contract. Unlisted units found on-site during routine service visits may be ignored, and with bad luck, a high-risk unit may remain in use without being serviced for a considerable period, this only coming to light when a fault or incident occurs. Conversely, equipment can languish unused but still be wastefully included on a service contract.

Even where parts and labour costs for repairs and maintenance may appear to be covered by a contract, a customer may, under some circumstances, still be expected to pay or even be refused a repair. Situations where support is excluded usually include misuse or damage beyond fair wear and tear. Accidental damage is a further category which may or may not be covered, and some contracts limit the number of events that will be accepted under such a provision, in what is effectively an accidental damage insurance scheme. It can be difficult to establish what constitutes fair wear and tear, but if it can be clarified at the outset, then difficult arguments can be avoided. Gross accidental damage may not count as fair wear and tear, for example, if equipment is crushed by being caught in a lift door, whereas minor damage from regular use can be expected. One way to resolve differences is to include provision in the contract for external arbitration, as is done with leasing contracts, where an expert acceptable to both parties assesses the state of the equipment against what might be expected for an item of similar age and application. *Misuse* is a general term that allows a supplier to withhold cover for the consequences of acting outside the manufacturer's specification – either using the equipment for an unintended purpose or looking after it incorrectly. For example, a supplier may claim that damage to reprocessed devices such as endoscopes was caused by inappropriate cleaning or disinfection. This situation can be avoided by obtaining manufacturer approval in advance for cleaning agents and processes, although it is difficult to manage where companies do not specify an approved procedure or only agree to support a single process that is not widely available. Each type of equipment can have its own specified ways in which actions by the user might void contract cover, and customers must check contract terms carefully when deciding how to set up ongoing support for their equipment. It is common for all contracts to stipulate that warranty or other support may be invalidated if the user performs unauthorised maintenance that involves, for example, breaking seals on the equipment or uses anything other than specified parts and consumables. There are often restrictions on how equipment can be used with or be connected to other devices. Storage conditions and operating temperatures are specified in the user manual, and although some manufacturers may be forgiving of slight breaches of these conditions, severe damage caused by obvious tampering, misuse or poor storage is unlikely to be overlooked.

### 9.2.1.13 Liability

The financial cost to an organisation of a medical device incident could be substantial, with potential legal liability for injury from equipment failing and injuring patients or staff or causing serious damage such as an electrical fire in a hospital. Product liability and consumer protection law generally apply to medical devices, and an injured person may be able to sue the manufacturer directly if equipment develops a fault due to inadequate design or manufacture. One defence is to show that the defect was not present in the equipment when it was sold but was acquired subsequently by misuse or inadequate maintenance. Under these circumstances, an organisation might seek to recover damages under their contract with a maintenance supplier, which is why the conditions set out in the contract are important. Providing an inadequate standard of support can make a maintenance supplier liable to pay damages, although contracts often seek to exclude liability for indirect losses, and it can also invalidate acceptance of future liability when switching maintenance suppliers, so customers must ensure any new maintenance provider accepts liability for previous work. Suppliers often charge for an initial service or assessment visit on a device they have not been covering previously.

### 9.2.1.14 Monitoring Measures for Quality of Service

Response, repair and downtimes are common measures which can be defined and specified in a contract. *Guaranteed uptime* is often quoted as the percentage of time for which equipment is available for use, excluding time out for breakdowns and routine maintenance, and is kept high by a combination of fast response times, short repair times, repair out of normal hours and substitution of working equipment. Statistics can be misleading and the definitions of *guaranteed* and *uptime* must be clarified. The end user seeks to have equipment operating for 100% of the time it is needed, but even a nominally high figure of 98% means equipment can be out of action 7 days a year without penalty, or even longer if downtime is measured over 24 h but occurs only during working hours. Failure by the service provider to meet performance standards specified in the contract results in some form of redress to the customer, often in the form of a payment related to the degree of performance failure. However, it is far better for the customer if a contract is delivered as specified. Financial penalties do not in general compensate effectively for loss of clinical service continuity.

### 9.2.1.15 Variations to Contract

Although most annual contracts will not require any change throughout the year and will terminate at the end of the contract period, changes or terminations may be necessary during the life of a contract, for example, when the number of devices covered alters, a whole fleet is changed or a contract is no longer economical. Customers should be aware of any penalties for premature termination, for example, when a multiyear deal has been negotiated but is being stopped early. A supplier is unlikely to refund money if the customer ends a contract prematurely unless the change is part of an ongoing deal tied to different business with the same supplier, such as the purchase of replacement equipment or a settlement for poor performance. Penalties can be extremely heavy where service contracts are bundled as part of an overall equipment management service, especially where agreements include replacement programmes outsourced to third parties. Under these circumstances, the provider expects to recover profits across the original lifetime of the contract, so if termination is more than a year early, any penalties can be sizeable.

## 9.2.2 Tenders and Negotiation

Tendering is an opportunity to compare servicing costs from the equipment provider against in-house and third-party options. Where this takes place as part of a new equipment purchase, the organisation holds a particularly good position to negotiate a servicing deal with equipment suppliers. It is more difficult to gain negotiating leverage when tendering for equipment support contracts alone, but there is potential to seek the best combination of maintenance support elements from multiple suppliers, as outlined previously. Throughout the process, the negotiating team will get the best outcome if it has the best information possible and maintains a united position.

Negotiations are usually led by representatives from procurement with the involvement of clinical engineering and the end user department, working through agreed channels and the organisation's appointed officers and committees. Each group brings its own expertise and experience to the table. Detailed knowledge is more effective than a tough approach by negotiators with limited background, and the use of *principled negotiation* techniques supports the achievement of a cooperative and positive outcome [1]. Whilst it can be tempting to talk directly to supplier representatives, it is important that all parties – end users in particular – do not do so outside formal negotiations. Attempts to bypass these processes, for example, by company representatives approaching staff directly, can result in premature disclosure of information and undermine negotiations, ultimately disadvantaging the organisation. Some organisations therefore have a policy requiring all company representatives to make initial contact only through certain people, though this is difficult to enforce. It is even more difficult when a potential provider of, for example, outsourced maintenance services contacts a senior manager in the hope of effecting a fait accompli before their proposals can be critically evaluated by equipment management specialists. The offer of instant returns to a finance director seeking savings can seem a very attractive proposition. Only later will any long-term overall loss to the organisation become apparent. Organisations should treat all such approaches consistently and review them as thoroughly as they would a formal tender.

Negotiation should reduce maintenance contract costs below manufacturer's list price for a given level of service, either because support can be sourced from suppliers other than the manufacturer or the organisation has a significant amount of business or other leverage with the supplier. Direct negotiations are one approach. Another possibility is to go through a consortium that procurement provides opportunities for discounts by negotiating bulk purchases of service contracts. However, it does not pay to assume bulk-negotiated prices are the lowest obtainable and there is nothing to stop individual hospitals negotiating an even lower price.

### 9.2.3 Operational Support

Cooperation between the organisation's staff and the external contractor is vital to reduce overall effort spent on maintenance. A number of key areas are highlighted in the succeeding text.

#### 9.2.3.1 In-House Technical Staff

In-house technical staff are able to provide small acts of support, including assistance out of hours, whilst service engineers can pass on tips about equipment operation and support. Where in-house technicians are providing front-line support, it may be beneficial for them to shadow external engineers, to improve their knowledge of equipment operation and learn solutions to common device problems, thus enabling them to reduce emergency call-outs and downtime. They can also assist the contractor with local knowledge and resources and speed repairs, for example, by helping external engineers find their way around the hospital or by receiving couriered spares from the manufacturer's distribution centre for the company engineer to pick up. Having in-house technicians at hand also makes it more likely that work is done to a good standard and allows service reports to be signed off with confidence.

### 9.2.3.2 Clinical User Involvement

Some end users welcome the involvement of the in-house clinical engineering service in service contracts, whilst others see it as a waste of time. It is important to strike a balance with the relevant departmental equipment manager. If departments such as imaging manage their own service schedules and call-outs entirely, it is the responsibility of their equipment manager to monitor and record work done and arrange for any necessary in-house support. Generally, specialist users tend to be more knowledgeable about their equipment and want to play a leading role in managing and supervising service engineers, as they appreciate the effects of downtime on their clinical operation. Clinical engineering staff play a more prominent role with ward and general equipment, where users are less interested in particular items.

### 9.2.3.3 Arranging Service Visits

For a contract to run smoothly, someone must liaise with external engineers to arrange access to equipment. The goal is to minimise clinical service disruption whilst making sure the maintenance programme is delivered. These responsibilities usually fall to lead equipment users or technical staff in the clinical engineering department. Procedures and responsibilities for day-to-day operation of the contract must be drawn up, and issues to be decided include the following:

- Who will organise routine maintenance visits?
- Who agrees the schedule of planned visits and makes them known locally?
- Who is present at routine maintenance visits to liaise with the company engineer, identify equipment and raise any detailed operational problems or faults?
- Who signs off satisfactory completion of routine maintenance visits?
- Who requests help for breakdown repairs? Is this via clinical engineering or a single contact in the first instance? Does the end user contact the supplier directly? Is authorisation required from a local clinical manager?

- Who authorises expenditure on visits/repairs outside the contract?
- Who signs off breakdown repairs as satisfactorily completed?
- Who checks and keeps records of these visits?

### 9.2.3.4 Labelling Equipment

Many types of equipment, medical or otherwise, may be present in a particular ward or department, and users need to know who to contact quickly in the event of equipment failure. All medical equipment should be clearly and visibly labelled with the contact details of who is responsible for maintenance and with its individual asset tag number. The distinction between equipment supported by different providers, and indeed between medical and other devices, may not necessarily be clear to the end user. Clinical engineering departments should be able to use their database and service contract records to answer or redirect enquiries.

### 9.2.3.5 Keeping Records

Users need to keep an eye on what work has been done under the contract and update equipment service histories. A prerequisite of this monitoring is to set up and maintain a well-indexed database of all contracts, allowing details to be accessed quickly. Records should cover the nature and scope of each contract and details of the company service department. Ongoing records of all activity relating to the contract can then be added to the organisation's own medical equipment inventory and equipment management database, including routine service visits, modifications and breakdowns. The design of the organisation's database should include a link between contract information and individual item service histories. This ongoing recording of information has three principal purposes:

- To help assess how appropriate it is to continue with the same contract type and level of cover in the light of ongoing equipment performance and reliability, changes in way equipment is used or the number of such items in the organisation.
- To monitor contract implementation and identify any problems getting access to equipment or performing routine services on time. Other less-evident shortcomings might include omitting key component changes or skipping checks, and these might only come to light when incidents occur or service providers change. In-house services sometimes see such problems when carrying out front-line support or quality assurance checks.
- To maintain an audit trail showing that appropriate maintenance and risk management has been carried out. This provides evidence for governance performance indicators.

### 9.2.4 Contract Performance Monitoring

Most contracts run smoothly, with no need to change arrangements during their lifetime. Performance monitoring will hopefully show the contract is appropriate and facilitate renewal at year end. In its broadest sense, contract monitoring aims to detect poor performance. This can be due to a failing by any party involved in executing the contract, whether contractor, in-house support service, local equipment manager or end user. The clinical engineering service should seek customer feedback on external contractors and on any in-house support to individual contracts and also ask contractors for their views. Unless all parties work together constructively, the system will perform more poorly than it should.

Contractors may, for a variety of reasons, not perform some or all scheduled routine service visits or maintenance procedures. Common reasons for this are lack of access to equipment due to clinical priorities or poor equipment tracking in the hospital. It is usual to blame a contract provider if a planned maintenance programme gets substantially behind, but it might be due to poor in-house processes or a lack of equipment, making what is there less available for service personnel to work on. Clinical engineering should, where possible, take an interest in service or breakdown repair visits and spend time liaising with company engineers and keeping an eye on the process. This helps an in-house service understand problems encountered by the company, give assistance in locating equipment and provide facilities to help cut service time and cost. Since delays or disruption to service visits can be caused by poor in-house practice, the number of service visits performed on time, whether by a contract provider or in-house, constitutes a key performance indicator for both equipment management processes and the maintenance service.

Monitoring may reveal that actual service operations are not as expected. Getting an end user's signature on a piece of paper is no guarantee that maintenance has actually been performed to specification by a company or contractor although, paradoxically, it may help as a defence against negligence on the organisation's part with regard to maintenance. Contractor problems with delivery may be due to a shortage of experienced company engineers, too high a volume of work being undertaken or lack of familiarity with a recently introduced type of equipment.

Statistics on the quality of service can be gleaned from the medical equipment management database, especially where it schedules and records preventative maintenance episodes and repairs. Two sets of statistics are relevant: those relating to a customer view of the service, which looks at time taken to solve the problem from start to finish; and those looking at the efficiency of in-house procedures. These are not necessarily the same. For an end user, the critical measure is duration of service is disruption. Where staff work shifts and there is frequent staff turnover, there is little opportunity to collect data about how well contracts are functioning. Responsibility for monitoring downtime usually falls to local equipment managers or the clinical engineering service and is often missed. This can be because *equipment downtime*, both planned and unplanned, is less relevant to users than *clinical service downtime*, so where replacements can be swapped readily for a faulty unit, there is little incentive for users to keep detailed records. Where areas depend on larger items of equipment, as in imaging, clinical service downtime is usually closely related to equipment downtime and so is critically observed.

Maintenance of performance monitoring data depends on good communication between end users, in-house personnel and company engineers. Data entry and validation of information are critical and rely on verbal and electronic communications and service reports. Paper service reports are liable to get lost unless responsibility is clearly assigned for signing and storing them – they can easily be thrown away or filed in a random pile by someone who does not understand their importance. If this happens, key opportunities for monitoring the service and recording information for liability purposes will be lost. One option is to store paper records in a defined location and another is to scan them into the electronic database. If scanned, quality system checks and records may be required to maintain an authentic legal record. An alternative is to require external service organisations to send in electronic copies of service records, either instead of or alongside paper copies, and this is a useful approach when monitoring service contract performance centrally.

It is just as important to keep records of in-house maintenance activities as it is of external contract providers, for the same reasons of equipment monitoring and requirement to show a duty of care. Manufacturer or multiservice third-party vendors will have clearly written contracts, whereas many in-house arrangements do not have a formal contract or recorded agreement. However, aspects of what is required may be implicit in the equipment management and risk management policies of the organisation, and operational arrangements should take full account of good practice. Customer feedback through questionnaires and interviews serves the dual role of monitoring internal system performance and satisfying the requirements of a quality system (see Chapter 14).

## 9.3 Contract Review and Renewal

Most maintenance contracts are renewed annually, but doing so automatically can lead to expensive inefficiencies and increased risk. Every contract should be reviewed before renewal and whenever poor performance is identified, to establish if contract type and level of cover are still appropriate. Reviews should gather and share feedback between equipment users, clinical engineering and the external provider to identify and rectify any shortcomings.

The first stages in reviewing a service contract are to check equipment covered by the contract is still in service, and that all equipment in service is covered. Users should go through the contract item by item to make sure equipment is still in active use and that serial numbers are recorded correctly. The way equipment is used clinically might also have changed, with newer equipment relegating items to a backup role. Are items likely to continue in service for the foreseeable future? Is replacement planned soon, and if so, can broken equipment be withdrawn and replaced early if it fails, rather than being mended? As when setting up the contract, users should be able to make a case stating the clinical value of equipment to the hospital and providing front-line experience of contingency arrangements, the impact of breakdowns on the service and highlighting any consequential costs. The review should also check for any shortcomings in contract implementation, by examining records of equipment performance and reliability including the number and type of user problems and breakdowns. Clinical engineering should have equipment history records from which reliability and the costs of in-house and ad hoc repairs can be estimated. Frequent breakdowns might indicate it is worth increasing the level of contract cover but may also indicate a need for imminent replacement. This information should be fed into the equipment replacement process along with associated costs.

To monitor and review contracts effectively, data need to be interpreted by staff with experience of equipment contracts and an understanding of finance. They should be good at negotiation and communication and have authority to be able to act on their findings. For most contracts, review will involve end users or clinical engineering, but large contracts should additionally involve procurement, finance and clinical service managers. If poor performance is identified, negotiation with the supplier and management action at a sufficiently high level will be needed to remedy it. The people involved can change through the lifetime of a contract, and good records of all communications between customer and supplier will help prevent confusion or misinterpretation of what has been discussed or agreed previously. A review will decide either to continue with the contract, terminate it or change it to a more appropriate one. If the last option is recommended, the organisation will have to renegotiate contract options, levels and prices with the company or go out to tender with another supplier. Where equipment breakdown history shows a contract is more or less cost-effective than previously thought, a short outline business case will help to persuade managers to change the way maintenance is approached for particular types of device.

Contract costs are a significant proportion of overall maintenance expenditure. Proportionate effort is called for to ensure they are managed efficiently. In a large organisation, contract management can occupy one or more people full time. If contracts have not been managed closely in the past, contract review should be a high priority, as it will be for anyone setting up or taking over an equipment management service, and it may well take months of work to establish whether all maintenance contracts are appropriate and equipment lists up to date.

## 9.4 Summary

Asking an external supplier to support medical equipment is one option to meet an organisation's obligations to make sure medical equipment is maintained appropriately. The customer still needs to have an in-house capability to be able to monitor and review contracts and ensure they deliver value for money. Partnership arrangements between in-house and external service suppliers can deliver a better quality of service than relying just on an external supplier. The development of longer-term relationships between customer and supplier can lead to an increasing element of good-will on the part of the provider, particularly where a customer spends a significant amount. This is helpful in smoothing out the inevitable hiccups that arise when trying to gain equipment support to meet clinical needs. Purchasers can encourage a positive relationship by meeting their obligations promptly and supporting external service engineers with appropriate facilities on-site.

## Reference

1. Fisher, R. and Ury, W. *Getting to Yes: Negotiating Agreement Without Giving In*, revised edn., Penguin Group, London, U.K., 2011.

# 10

## Adverse Incidents, Investigations, Control and Monitoring

### 10.1 Introduction

Understanding how and why an event happened is the first step towards reducing the rate and impact of accidents. Most accidents can be prevented by doing something differently. Taking a structured approach to an incident investigation makes it much more likely that root causes will be identified and corrected, rather than simply blaming a problem on the actions of individuals. The practice of incident analysis and continual improvement has been behind the remarkable reduction in accident rates achieved in aviation and construction over the past decades. Medicine has a long way to go to reduce rates anywhere close to those achieved in these high-risk industries, and background figures showing this are presented in Chapter 3.

   In this chapter, we describe the concepts and practices associated with adverse incidents and their investigation. We define ways to categorise incident severity and describe the chain of events that follow a medical device incident, from establishing the actions of end users at the time an event occurred to making changes to practice and procedure to prevent recurrence.

### 10.2 Definitions and Categories

The World Health Organisation defines an adverse incident as 'An injury related to medical management' [1]. This definition can be widened to include events putting the organisation at risk. An important subcategory is the serious untoward incident (SUI), one that causes unexpected major injury or death to a patient, visitor or staff or has other serious implications for the healthcare organisation. Judging an untoward incident as serious will automatically require a more intensive and thorough investigation and involve higher-level managers, both important when action needs to be taken quickly to retain critical evidence and limit negative consequences. One way to provide additional guidance on what constitutes *major injury* and *serious implications* is to grade incidents using the quantitative risk management process described in Chapter 3. Because no guidance can unambiguously determine whether an incident falls into one category rather than another, senior staff members need to be involved in classifying incidents as *severe*.

A second, separate incident category is the *near miss*, which is an event that could have led to an incident happening. This definition is intentionally very wide. Paying attention to near misses, and seeking to prevent them occurring again, is proven to improve safety in critical areas and reduce the frequency and severity of any incidents that do occur, as described earlier in Chapter 3. Healthcare organisations in the United Kingdom are encouraged to develop a *no-blame* culture to incentivise staff to identify and report near misses and be proactive in preventing any recurrence.

Incident reports are collected and published by various national and international bodies and independent organisations, but there are difficulties comparing and analysing incidents due to the lack of a universally agreed classification. Overarching bodies such as the ECRI Institute and national safety bodies have developed their own schemes for classifying adverse incidents, but these are not universally accepted, and individual organisations and analysts often develop their own. This makes it difficult to compare incident rates, a problem compounded by chronic under-reporting as discussed in Chapter 3.

## 10.3  Why Report Adverse Incidents?

At first sight, it might seem desirable to project the image of an organisation in which nothing ever goes wrong. Yet in the long term, this is never true – problems and difficulties will always arise. When they do, an organisation that has denied problems exist is likely to fare far worse than one which faces up to and seeks to improve its faults. So paradoxically, an organisation which reports no incidents is unlikely to inspire confidence in its safety record but rather to raise suspicions that its risk reporting is substandard.

The primary reason for reporting incidents is to prevent further incidents of the same kind happening again, both within the healthcare organisation and in the wider healthcare community. Reporting may also be required to meet legal obligations and governance requirements. Even where reporting is not mandatory, it shows that processes are in place to manage risk, bolstering any defence against litigation. Investigations help organisations manage complaints and freedom of information requests and disseminate valuable knowledge to manufacturers, other organisations and the public. Where legislation does create a duty to report, it usually defines what should be reported and identifies the notified body to which a report must be made (see Chapter 3). An employer who fails to take reasonable care to protect an employee from foreseeable injury can face legal action, potentially being sued for damages by the employee and prosecuted under criminal law for lack of care. Effective incident reporting and management systems are evidence that the employer has taken reasonable care. Detailed records will show how the incident arose and why the organisation responded as it did and are likely to prove helpful in any subsequent litigation or prosecution.

Even when reporting is not a legal requirement, it can help others improve safety by building up a learning database which can be used to identify and reduce risks to patients. Voluntary reporting can occur at national or international level and is likely to be targeted at particular risk areas, such as the anonymised European Society for Radiotherapy and Oncology (ESTRO), European database for recording incidents in radiotherapy [2]. Healthcare organisations can use these databases as a resource to help improve their clinical and organisation practice and reduce the incidence of adverse events.

## 10.4 Initial Incident Handling

All incidents and near misses should be investigated and reported on as soon as possible. Organisations should set out a procedure for how this is to be done that covers both clinical and medical device incidents, as these are often interrelated, and includes clear directions and flowcharts showing how incidents should be handled and reported. Staff should be made aware of the need to report incidents as part of their safety training and be reminded regularly of detailed policies and procedures relevant to their workplace.

Action in the initial period after an incident occurs is critical if an accurate record of events is to be obtained whilst memories are fresh and before evidence is removed or discarded. Actions to take when an incident or near miss is first recognised include the following (see Table 10.1):

- Quickly assess immediate risks and make the local situation safe.
- Ensure continuity of clinical treatment as necessary, for example, by providing alternative equipment or an alternative clinical procedure.
- Quarantine any devices involved to prevent further use or tampering and preserve evidence by maintaining equipment settings at their existing value or recording settings if safe to do so.

**TABLE 10.1**

Initial Handling of a Device-Related Incident: Summary

| Steps to Take When Initially Handling a Medical Device–Related Incident |
| --- |
| 1. Make immediate situation safe. |
| 2. Assess initial seriousness; contact senior managers if concerned. |
| 3. Take wider immediate action if required (suspension of service, recalls, external notification). |
| 4. Note all device details and settings (including make, model and serial/batch numbers). |
| 5. Retain and quarantine device/consumables/packaging involved where possible. |
| 6. Record who has been notified (MHRA, service personnel, manufacturer, others). |
| 7. Write initial incident report. |

- Keep associated consumables or accessories, with due regard for any infection control implications, together with any packaging, labels or documentation containing batch or other reference numbers.

- Notify appropriate internal bodies where immediate or short-term action is required, including clinical engineering and healthcare professionals dealing with the patient.

- Complete the local incident report pro forma and submit it through appropriate risk management channels.

Initial reporting is likely to be made to local managers or team leaders in the department or service where the incident took place. The incident handling procedure will contain guidance on how to assess the severity and consequences of an incident and judge its associated risk, and depending on its nature and severity, it may also be reported internally to the risk management department of the organisation and possibly externally, for example, to medical device or health and safety regulators. Lower severity incidents are likely to be addressed and managed locally whilst more severe events are likely to be reported initially to senior management, the organisation's duty risk manager and possibly executive directors. The turnaround time for an initial report will depend on risk level and typically ranges from immediate for high-risk incidents to consideration at weekly or monthly meetings for low-risk incidents and near misses.

Certain incidents involving medical equipment should or must be reported to a notified body, as discussed earlier. However, the range of events an organisation will report depends both on national legislative requirements and guidance, and on how the organisation interprets them. Some organisations are reluctant to report incidents externally unless they have to, through fear of regulatory action, so the attitude of regulators has a direct influence on openness. Discretionary reporting also raises concerns about indirect censure, which is why anonymised systems are more likely to attract reports. In one study, anonymous reporting increased the proportion of reports showing operator error from 50%, typical of that found in national reporting systems, to 80% [3]. However, the maximum value is extracted from each incident if it can be thoroughly investigated, and for this reason, attributable reporting is implicit in the Medicines and Healthcare products Regulatory Agency (MHRA) monitoring system and in risk management mechanisms within hospitals. Such systems need to operate under a *no-blame* culture, where staff members do not fear repercussions, to maximise the number of reports and investigation accuracy. The UK MHRA encourages reporting of any device-related incident that, 'causes, or has the potential to cause, unexpected or unwanted effects involving the safety of device users (including patients) or other persons' [4] and expects all adverse incidents involving medical devices to be reported within 24 h or less. Of approximately 10,000 events reported to the MHRA in 2010, 40% were noted and used in trend analysis, 40% led to a standard investigation and 20% triggered a detailed enquiry.

Many equipment breakdowns or malfunctions do not constitute incidents requiring investigation or reporting unless they highlight repeated failure of particular components or present a potential danger to patients. Thus failure of a pulse oximeter finger probe after mechanical damage or an isolated battery failure in a syringe driver after months in operation is to be expected and would be of interest only as a statistic on the internal equipment management database or as examples where a department did not have equipment fit for purpose. In contrast, consistent premature battery failure in a particular model of syringe driver is a cause for wider concern and should be reported. The device manufacturer may also need to be informed of the incident, and the notified body may specify requirements for a manufacturer investigation.

## 10.5 Incident Investigation

### 10.5.1 Investigation Process

The purpose of an investigation is initially to uncover why an event happened, as illustrated in Figure 10.1. For minor device incidents managed and analysed within the clinical engineering service, investigations are

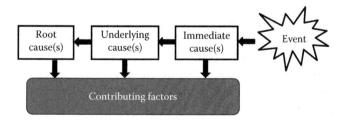

**FIGURE 10.1**
The investigation – working back from the event to uncover why it happened.

**TABLE 10.2**

Summary of Steps to Take When Carrying Out a Root Cause Analysis

**Summary Steps of a Root Cause Analysis**

1. Identify the investigator/investigating team.
2. Establish a detailed chronology.
3. Analyse – what happened that was not designed to happen.
4. Identify contributory factors and for each one identify immediate and underlying causes.
5. Assess controls – for each factor/problem, identify and assess existing risk control measures and estimate the difference each control measure makes using a risk evaluation scale.
6. Report – sequence of events, interpretation, identified risks and recommended ways to reduce these to an acceptable level.

ideally undertaken by one or two people knowledgeable about equipment and clinical operations in the area where the incident occurred. Incident management may simply consist of notifying a non-conformance through the quality system or modifying equipment support and maintenance procedures. In-house investigation of more serious incidents may involve a specially convened working party of clinical and technical experts and include the healthcare organisation's risk management team. This group will make decisions on the nature of the investigation, delegation of subtasks, level of reporting and production of reports and will devise and consult on the implementation of new or modified procedures to prevent reoccurrence.

The most effective incident investigations follow a structured process. The technique of root cause analysis looks beyond the immediate cause of an incident, examining its structural, environmental and procedural context to uncover factors, often long term, which created a situation where the event could occur. The main steps in a root cause analysis investigation are set out in Table 10.2.

## 10.5.2 Establishing the Facts

There are typically four elements to this part of a medical device incident investigation:

1. Examine written incident reports, such as the standard pro forma and any other written records or correspondence relating to it, including user training records.
2. Interview eyewitnesses, referring to and cross-checking independent statements where possible.
3. Examine the device and associated equipment, looking at any relevant records or device logs.
4. Perform any appropriate tests to diagnose what may have happened.

**TABLE 10.3**

Suggested Approach to Take When Interviewing Staff

**Key Points When Interviewing Individual Staff**

1. Explain the purpose of interview and role.
2. Stick to the facts, avoid emotions and jumping to conclusions.
3. Listen actively and sensitively.
4. Minimise stress – ask easy questions first and avoid leading questions.
5. Construct a detailed individual chronology – verbally first and then interactively in writing.

There is a distinct skill to carrying out eyewitness interviews, and Table 10.3 highlights a useful pattern to follow. A good method of getting started is to ask open questions such as, 'After that, what did you do next?' rather than closed questions such as, 'Did you then press the red button?' Closed questions are useful however to confirm understanding or help an interviewee recollect what they said earlier. The investigator should prepare for the interview by reading any specific and general procedures and policies that should have been followed. Questioning can then establish whether or not these were adhered to. All users are responsible for carrying out pre-use checks on equipment, ranging from a quick examination for damage to a detailed and recorded sequence of tests, so the investigator should examine available records and other evidence such as quality control samples for evidence of whether necessary checks were carried out and then verify this at interview. Information obtained directly from the medical device and associated equipment can be cross-checked against witness testimony, a useful way to establish the completeness of an individual's view and understanding of the events that occurred. The investigation will establish if the end users were trained, competent and authorised to use the device (or were adequately supervised) when carrying out the procedure and whether or not their training was up to date regarding any device or procedural changes.

### 10.5.3 Investigating the Medical Device and Associated Equipment

The medical device and any associated equipment must be examined and tested for any damage or malfunction. This may require specialist investigation, for example, looking at the quality of material and production processes in the case of a consumable failure. For medical equipment, safety and functional tests include setting controls to the values they were reported to be at when the incident occurred, if this is practicable and safe. Specialised equipment may be needed to mimic conditions under which the incident happened, using available test gear or phantoms or building one-off test rigs. The extent to which this is required will depend on the severity of the incident and, for a serious event, can evolve into a significant investigation in its own right. Many life-critical devices and more complex computer-based equipment such as anaesthetic machines, ventilators, critical care grade infusion devices and imaging equipment have internal event logs which record detail about all user operations and clinical procedures carried out. This log is a critical piece of evidence and can be interrogated to show the exact sequence of settings leading up to the incident. Examination and testing will be carried out with the original device or similar interconnected or interdependent equipment and consumables.

If environmental factors are involved, for example, with a suspicion of possible electromagnetic interference or interaction with other equipment, then any tests should be carried out as far as possible under the same conditions. It may be necessary to examine and test other equipment or infrastructure such as the electricity supply. Factors such as the integrity of electrical isolation, earthing, over-current protection and compliance with wiring regulations should also be looked at. Device and equipment tests may be carried out by the in-house service, the manufacturer, external consultants or any combination of these. For serious incidents, the relevant regulator may want to be involved in verifying the effectiveness and integrity of any investigation. Independent testing or assessment avoids any accusation of bias by either the equipment user or the manufacturer and is helpful if significant litigation around the event is possible, and notified bodies should keep a list of suitable reference organisations.

An investigator will also want to examine the database record for any medical equipment involved will be examined, looking for any previously reported faults and incidents and checking for any hazard notices relating to the device. The investigation will be concerned to establish whether any pre-existing condition or series of failures might have contributed to the incident, including inadequate maintenance. The investigator should also identify any recent changes to best practice, both through manufacturer bulletins and in consultation with relevant professional bodies, and see whether these have been implemented.

## 10.6 Incident Analysis

Having obtained all accounts and reports of the incident and assembled background information, the analysis phase combines and compares the evidence to identify all contributory causes to the incident. This gives the investigating team the basic material to devise appropriate remedies and control measures. Three broad categories of incident causes are human error, system failure and equipment failure. The boundaries between these areas are indistinct, and there may be several contributory causes to an incident rather than one single failure alone. As discussed earlier, a blame culture seeks to identify individual human error as the cause, but many incidents can be prevented through better control systems, and equipment failure in use may be due primarily to poor training or equipment management. Investigators need to avoid premature judgements when carrying out an analysis and to think as broadly as possible when looking for potential causes.

### 10.6.1 Investigation Techniques

A general knowledge of previous incidents, both within the organisation and outside, is helpful in both investigation and analysis phases. An experienced investigator is likely to be familiar with similar previous problems and can guide the investigation along profitable lines. Investigators can keep up to date by reading reports, digests and bulletins published by government bodies, looking at books on medical devices [5] and risk management and from reports and meetings of professional organisations [6].

Background knowledge can be very valuable but experience is only a guide. Investigators should remain open to new possibilities, as totally unexpected incidents with highly idiosyncratic causes can occur at any time. To be thorough and make sure no possible cause is overlooked, an investigator must systematically examine every aspect of an incident. One useful technique is the *fishbone* diagram (see Figure 10.2), which links possible contributory causes in a framework to guide investigation. These areas are explored in more detail later, and the diagram provides a way to summarise the results of an enquiry. A second technique is *Five Whys*, which aims to go beyond immediate causes and drills down up to five levels by repeatedly asking the question, 'Why did this occur?' So once an initial cause is established, the reasons for that cause itself are investigated, and so on. The following is an example:

Incident: No ward staff available to answer patient calls for assistance.

Why? Duty Sister had gone to look for BP cuff on another ward.

Why? No spare was found on the ward.

Why? Did not know where spares were stored or how to order them.

Why? New to unit and no local induction.

Why? Unclear responsibilities and lack of time.

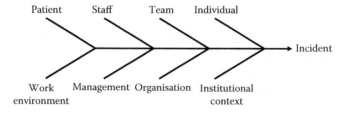

**FIGURE 10.2**
The *fishbone* diagram to help identify contributory causes of an incident.

This example uncovers weaknesses directly contributing to the incident, including the staff member leaving the ward and a lack of spare blood pressure cuffs, and others of more general significance that could potentially contribute to many different kinds of incident. A lack of induction in ward procedures due to pressure of work could lead to inappropriate actions in a wide range of instances. Further investigation may explore issues such as ward staffing levels, attitudes towards training, relationships between staff on the ward, and support from supplies and clinical engineering. It may also uncover a lack of responsibility for spares management.

### 10.6.2 Human Error versus System Failure

Human error is variously reported as being the major cause of between 15% and 80% of incidents. This surprisingly wide range arises from the interplay of two different viewpoints adopted by analysts looking at incidents where human error was involved: the person-centred approach and the systems approach. The former attributes the primary cause of events to human failings, when these are a causal factor, and seeks to explore how such failings may be prevented in the future, through changes to procedures and retraining of individuals. The systems approach considers human error one factor amongst many when identifying and interpreting root causes, with the investigation and subsequent recommendations aiming to improve the system and manage its risk to make it more resilient to individual human error. In reality, there is a continuum between these two extremes. Human error might be due to carelessness, which in turn might have been induced by tiredness due to unrealistic working patterns, something which could be considered a *system error*. When a system does not foresee common events such as an urgent need to provide sickness cover, or guard against likely human errors arising from these situations, then the system rather than the individual is at fault and needs to be redesigned.

Malfunction of a medical device is classed as equipment failure. Analysing the root cause and trying to prevent similar true equipment failures would reduce but not eliminate repeat events, as unforeseen equipment failure is always a possibility. The real clinical damage is usually caused by lack of effective contingency and backup arrangements, so improving these systems is likely to be the most effective way to minimise the future clinical impact of any malfunction.

Whether the primary cause of an incident is human error or equipment failure, remedies suggested at the end of the investigation should address all root causes identified. Long-term harm reduction relies on improving system design and implementation, and investigations that take account of the findings from similar and related incident enquiries are likely to come up with more effective recommendations. This is particularly true when considering how to avoid human error, as the study of behaviour under different conditions continues to highlight how the frequency and consequence of mistakes can be reduced [7].

### 10.6.3 Categories of Human Error and Associated Remedies

Human errors fall into a number of interrelated categories. Basic mistakes are usually made more likely by ergonomic errors in the design of equipment and systems, pressures of work or culture, a lack of training and individual behavioural problems. Each of these is considered later, and final paragraphs comment on ways to reduce errors and encourage positive behaviours.

#### 10.6.3.1 Poor Design

Poor design of a machine or system makes operator error more likely and exacerbates the causes of reduced concentration, outlined earlier. Ergonomics and human factors engineering seek to better match the task and a worker's physical, information handling and workload capabilities [8]. Ergonomics is the science of maximising the efficiency, clarity and ease of use of the human/system interface. The distinction between human and system error here is particularly blurred. A logical and consistent device control layout, together with clear, legible and unambiguous labelling, leads to ease of operation and reduced operator stress. Standardising associations between device control functions and their labelling, layout, colour, etc. will reduce the risk of accidents – as occurred with manufacturer standardisation of clutch, brake and accelerator positions in automobiles. Confusion is particularly likely when device control positions, units of measurement, marking or operation vary between different models of otherwise similar devices. Previous experience with equipment from another manufacturer designed to an alternative set of conventions, for example, the positioning of rotary knobs controlling different gases on an anaesthetic machine, may lead to errors when ingrained actions or *automaticity* takes over from conscious performance [3]. In order to prevent such confusion, international standards for particular device types may well specify the shape, marking and direction of the operation of particular device controls but cannot cover all eventualities. Users must be retrained whenever there is a change in technology and especially when a mixture of old and new technology is in use simultaneously, with relevant instructions and advice available. A particular example is the change from monophasic to biphasic defibrillators, where energy levels used in the two technologies are different and confusion has caused resuscitation failures [3]. Clear summary instructions are essential, as users will be unable to refer to a complex and comprehensive user manual in a clinical emergency, and the mantra, 'If all else fails read the instructions', will usually come too late. Lack of standardisation can lead to operator overload, as happens when multiple alarms sound on a variety of patient monitors or infusion devices in high dependency areas, making it easier for users to ignore alarms than spend time working out what has caused them in the first place [9].

### 10.6.3.2  Basic Mistakes

Mistakes can be of the simplest kind, such as mixing up right and left or clockwise and anticlockwise. More subtly, a user may be confused by controls which look the same but have different functions, a particular problem when switching between equipment models or changing interface settings in software. Examples of basic errors include plugging ECG electrodes into a floating mains connector and plug [3] and commencing cardiac resuscitation when an ECG monitor displayed a flat line due to disconnected electrodes. Such incidents are now guarded against by a changed plug design and lead-disconnected alarms, respectively. Another possible error is to select a moving ECG trace stored from a previous patient rather than observe that from the current one [10]. In making these kinds of mistakes, the operator does what seems to them to be correct. If this leads to an unexpected outcome, the operator may persistently retry the same action in the usually mistaken belief that events will correct themselves, rather than stepping back and asking why the apparent anomaly has arisen. An example of a situation like this where common sense appears to have been ignored was when staff persistently believed the erroneously high reading from their local portable glucose meter, despite clinical signs and a laboratory sample analysis showing the patient was hypoglycaemic, delaying vital treatment. To an onlooker, mistakes such as these are hard to explain. From an operator's point of view, however, factors such as tiredness and stress may impair rational mental function and lead to a lack of concentration or inability to cope effectively with distractions from interruptions or questions, including mobile devices [11]. These factors can easily precipitate disaster, especially when coupled with lack of training and unfamiliarity with a particular model of equipment. Tired people are particularly prone to guessing or taking chances, hence the tendency to base assumptions about the operation of device controls on experience with other models. Unclear or ambiguous instructions can also be interpreted or applied incorrectly if a user is not fully alert.

### 10.6.3.3  Lack of Knowledge

Lack of background knowledge and understanding, can lead to serious consequences, particularly if coupled with overconfidence or misunderstanding and a lack of expert supervision. There are cases of radiation overdose in CT scanning where clinical staff have not appreciated the level of radiation dose to the patient as being several hundred times the level of a standard x-ray and have consequently performed multiple scans with very little clinical justification [12]. Also without specific training, a healthcare worker may not realise a test sample requires special handling or storage before being sent to a laboratory for results to be meaningful, leading to erroneous results, misdiagnosis and serious potential consequences. Such operator errors might be avoided by better training, a system of expert overview and improved protocols.

### 10.6.3.4 Lack of Training and Experience

Lack of training and experience amplifies any human or ergonomic weaknesses in a system. Procedures involving medical devices usually involve attention to a number of factors such as observation of and interaction with the patient, monitoring indicators and device control settings, and operating device controls. Lack of familiarity with any of these factors reduces user concentration on the others. If an operator is unfamiliar with a procedure and does not check that the quantity being measured or delivered is realistic, then the wrong action may result. There are numerous examples of maternal heart rate traces being mistaken for foetal traces during foetal monitoring, including cases where the foetal signal was absent entirely. These signals should be differentiated by their mean heart rates, but if an operator does not check the chart scales or is unfamiliar with expected values, then confusion can occur [3]. Also if accessories vary in form or size to suit different types of patients, then choosing the wrong size can cause erroneous readings with, for example, patients being treated inappropriately for low blood pressure because a wrong sized cuff gave too low a reading [3]. Many incidents have been caused by impromptu repairs or device substitutions carried out by people not qualified to do so. Problems arise when individuals without sufficient competence to undertake a task also lack the judgement to realise both the extent of their own incompetence and when they need to stop and get help [13].

### 10.6.3.5 Malicious Acts

Malicious acts such as sabotage and arson are sometimes carried out in the healthcare sector. Motives include mental illness, political beliefs, revenge and animosity towards the organisation or a desire to incriminate or harm colleagues. Some acts are deliberately designed to harm patients [14]. Actions may be disingenuous, where employees refuse to carry out a procedure that violates a trivial risk management rule or maintain they were not told exactly what to do, so as to embarrass their managers.

### 10.6.3.6 Behavioural Problems

Behavioural problems include negligence and alienation from the job, team or institution. The individual does not seek to optimise their performance, or that of others, and may fail to inspect systems before use or pass complex tasks to staff unqualified to do them. Individuals may take deliberate risks by wantonly disregarding safety features in order to save time, effort or annoyance, disabling alarms or removing safety guards. Drug and alcohol abuse also cause incidents and lead to low self-esteem and poor working relationships.

### 10.6.3.7 Institutional Pressures

Institutional pressures can lead people to perform less well. Factors such as a high workload and tight timescales lead individuals to attempt to complete tasks unduly quickly. Such an environment exacerbates any weaknesses in system processes and makes it more likely errors will occur. Heavy handed management will in time degrade human relationships and lead to a situation where individuals will not only be less likely to report near misses and errors but will also take risks and shortcut procedures in order to get the job done, with workers colluding to hide problems from their managers. Organisations wanting to encourage positive behaviour and increase incident reporting rates can do so by moving away from attributing incidents to personal human error – otherwise staff are more likely to withhold vital information and to shift the blame from themselves to others or equipment, obfuscating the true cause of an incident. Equipment may be sent for repair to cover up a lack of training or understanding, wasting time on unnecessary investigations and tests; fictional equipment malfunctions may be reported; or damaged equipment may be quietly shelved and its role in an incident be hidden. A blame culture also makes individuals fearful of admitting they do not understand, leading them to cascade inaccurate information to other staff.

What is the role of individual responsibility? In many cases, ways of minimising human error lie with the system and not the individual. A fair and open culture has the goal of establishing all the causes of an incident and of getting individuals to accept responsibility for their own actions, with an action plan to prevent any repetition of a similar or related situation. Forgiving genuine mistakes and taking measures to avoid their being repeated are more effective in the long run than censuring staff whenever they make a mistake. Managers have responsibility for ensuring individuals are trained for each task and are supported to do it. Effective management of excess workload leads to the development of sustainable working patterns, reducing stress and improving morale. Ergonomic system design and standardisation lessen the impact of tiredness and operator confusion. However, where it is not possible to alter individual behaviour effectively because of their own cognitive or behavioural limitations, removing the individual from the role may be the only safe course of action.

### 10.6.4 Equipment Failure

In addition to the obvious case of complete breakdown, there are many other ways in which equipment failure can compromise clinical processes. Subtle faults, miscalibration or misuse can produce misleading measurements or deliver incorrect treatment. Equipment can be used for applications or on patient groups for which it was not intended, or it may become technologically obsolete and fall short of current performance or safety standards. All of these possibilities need to be considered in an incident investigation.

Intermittent faults are most likely to occur in equipment that is old and unreliable or poorly maintained. Unreliability for some types of equipment increases with age and depends on its components, level of use and maintenance support. Understanding the ageing characteristics of equipment involved in an incident will help in the investigation but cannot rule out concealed damage or atypical performance. Recognising reduced equipment reliability and arranging replacement at an appropriate time will reduce the number of incidents related to equipment failure.

Subtle equipment faults may only show themselves at some control settings, combinations of control settings or after particular event sequences. With many variations of controls possible, and the use of software control systems in most devices, undetected faults may be present either due to inadequate warnings or because it is impossible to test systems software under every combination or sequence of settings and inputs/outputs. Such faults have sometimes resulted in excessive outputs of radiation, for example, when x-ray equipment has been left in a service mode allowing users inadvertently to change settings that are normally fixed. It is often extremely difficult to trace intermittent and subtle faults [3,15].

Equipment should be adequately checked, tested, calibrated and subject to regular and appropriate quality assurance. For example, every operator has a duty to check equipment for obvious faults before use, and not doing so is both a human and a system failure. Also without regular checks, accumulated changes and subtle errors can lead to clinical mismanagement, such as a missed infection in an infant due to a misreading electronic ear thermometer with a dirty tip [3]. Where regular replacement of seals or moving parts is a maintenance requirement critical to continued operation, failure to carry it out will almost certainly result in equipment malfunction sooner or later. Many faults can be prevented by regular inspection, and risks to patients or staff from electrical shock are often due to obvious faults which could have been detected by simple inspection (see Appendix B). Mechanical problems such as the collapse of hoists or failure of ceiling supports can also be prevented in this way. The thoroughness of such checks needs investigating in any incident.

When looking for causes of alleged equipment failure, a useful rule of thumb is that a third of reported equipment faults are due to poor user training or misuse, another third to minor faults from consumables or accessories such as leads and batteries and the remainder to breakdown or failure of the equipment. To operate equipment reliably, it is essential that adequate equipment spares and consumables are available, and local stores should be examined to check that this is the case. Equipment should be kept in a fit state for use, a particular problem with battery-powered infusion devices which are particularly prone to fail during patient transport due to inadequate procedures for battery charging [3].

The device types most likely to be involved in incidents are revealed in reported statistics. The most consistent top three are wheeled devices, surgical equipment including consumables and implants, and infusion devices including those for transfusion and dialysis [4]. Knowing which devices are most at risk in a healthcare organisation helps in planning risk and control measures, including proactive device checks. It is not just equipment that needs checking, and faulty accessories have been found including new tubes containing small pieces of debris that could cause blockages or be ingested by the patient. The low frequency of such events makes them difficult to guard against, as staff automatically assume these components are fine. An important lesson for incident investigations is to make sure all accessories, packaging and other items are retained and looked at carefully to see whether they could have contributed to or caused an incident.

Because multiple equipment items or accessories can be connected together inappropriately or incorrectly, it is important that novel or unusual equipment configurations are checked and approved by competent clinical and technical staff. Even if working to a standard configuration, it is wise to check all systems are connected, set up and working correctly both initially and from time to time, particularly where multiple devices are connected to a single patient. This is the reason why the overall equipment layout should be preserved or noted carefully for investigation, so that the same items can be used to recreate the environment in which the incident occurred. Otherwise, there is little chance of picking up a problem caused by the interaction of two independent devices, where two subtle faults may interact.

In summary, there are many reasons why equipment does not perform as intended. Root causes often include failure of the operator or responsible user to act appropriately at some point. However, human actions take place within systems which themselves have a significant impact on the human operator and also on the likelihood and impact of equipment failure.

### 10.6.5 Systematic Interpretations and Associated Remedies

Incidents happen in a context. Whatever their primary cause – human action, equipment failure or some other reason – many other factors influence the likelihood of an event occurring and affect its severity. The answer to preventing or minimising the impact of incidents therefore lies in changing the whole system, defined as the environment, procedures and interconnected processes within which the event occurred. Changes can be small and focused, such as changing maintenance arrangements from nominal annual inspection to mandatory proactive replacement of key parts, or broad and sweeping with many changes implemented simultaneously. In practice, incremental and sustained improvement across the whole system is the best way to reduce the severity and frequency of events, with careful attention to detail and a mindset that looks for every opportunity to improve.

A systems approach takes human errors, equipment failures and other untoward occurrences as inevitable. It designs the system to keep these low in number and contain their consequences, grouping factors into related

**TABLE 10.4**

Factors Affecting a Medical Device Operator Which Can Be Considered When Investigating an Incident

| Patient | Unexpected clinical change | Unintended clinical consequences | |
|---|---|---|---|
| Environment | Knowledge, training and skills | Interruptions, queries | Hazards |
| Procedures | Unclear procedures | Conflicting requirements | |
| Control measures | Control access | Limit actions | Require confirmation |
| Equipment factors | Unexpected operational problems | Unintended treatment consequences | Lack of consumables or accessories |
| Human factors | Knowledge, training and skills | Distraction, tiredness etc. | |

classes as depicted in Table 10.4. Considering these systems factors helps to identify possible incident causes, devise control measures and define their properties to suit particular activities and processes. The following paragraphs outline and comment on these individual elements.

### 10.6.5.1 Patient Factors

A patient may present with specific medical conditions, including behavioural, comprehension or communication problems, which affect the use of equipment or present additional hazards. An ideal system will identify and assess the risk associated with these and then modify procedures as appropriate, bringing in additional specialised equipment or staff. For example, heavy patients may place excessive mechanical loads on trolleys and couches or not fit into standard scanners. Maximum loading weights and dimensional handling capability of medical equipment need to be established beforehand to allow contingency arrangements to be devised or appropriate equipment purchased and any manual handling assistance arranged. Behavioural problems can require specially trained staff, the presence of relatives at procedures or the use of sedation or counselling. Patients may, if confused or disorientated, attempt to adjust, disconnect or dismantle nearby equipment. Such instances include a patient connecting an oxygen tube to their own intravenous line [3]. The systems solution here is to identify patients at risk and provide appropriate supervision and special care regarding the placement and connection of suitable devices.

### 10.6.5.2 Environmental Factors

Good health and safety management, coupled with analysis and risk assessment of working conditions and the working environment, can identify accidents waiting to happen. In environments populated with medical devices, poor layout, unstable stacking, trailing cables and a generally cluttered work environment can lead to damaged equipment and staff slips, trips and falls. More subtly, a cluttered visual environment will make it more difficult to see if an item is missing or an infusion wrongly set up. Systems should be in place to monitor and improve the workplace environment, and constant discipline is needed to keep on top of potential chaos in acute areas with fast patient turnover. Simple measures such as structured storage or equipment docking units make it easier to see if equipment is missing and provide a good design solution for multiple devices. Building in user reminders and regularly reviewing processes help to keep staff aware of potential dangers. Separating equipment from food and drink removes a possible electrical hazard, although spilt liquids will still create slippery surfaces and potential slip hazards.

### 10.6.5.3 Process and Equipment Factors

Principles of good equipment management and safe working practices should be built into the organisation's policies and procedures. These are delivered in turn through detailed protocols and their intelligent interpretation in a particular context. Systems may end up becoming unsafe, either because the underlying decisions on which they were based are conceptually flawed or due to improper implementation. Any investigation to detect causes of system failure should work backwards to find out if the principles of good equipment management were followed and, if not, consider where and how the concepts, implementation or both came to be lacking. Procedures and protocols should be clearly written and capable of straightforward implementation. For example, a detailed decontamination form for use when returning equipment for repair can be replaced by a simple sticker to show an item has been cleaned.

### 10.6.5.4 Operator Factors

In order to perform effectively and safely, an operator must work within the limits of their practice and competence. This includes recognising their own responsibility to anticipate and prevent problems and recognising they may be prone to making mistakes. Because people find it difficult to judge their own competence and mental state accurately and are prone to wishful thinking, systems should stop individuals being pushed beyond the limits at which they cease to act safely. Supervisors and managers should also ensure operators are resourced, trained and supported and provided with sufficient information. Systems which are fit for purpose provide guidance, training and supervision to equipment operators, including accessible instructions and expert advice in case of queries. A clinician working with patients and multiple devices faces what can be a complex and confusing set of controls and interconnections, making it easier to set up devices and accessories wrongly. Systems solutions include manufacturing connectors to different designs, so that, for example, arterial and venous infusion lines cannot be interchanged [3]. With multiple life-critical devices, such as in an operating theatre, alarms may be difficult to distinguish from each other and may be silenced by an operator in order to avoid distraction, leading to important alerts being missed if a patient later becomes compromised.

### 10.6.5.5 Team Factors

The overwhelming majority of healthcare procedures and interventions are delivered by teams. A patient visit to accident and emergency is likely to involve dozens of individuals from different professions working together to diagnose, prescribe and deliver care, backed up by many more delivering the physical, organisational and governance infrastructure. Under these circumstances, clear allocation of responsibility and good communication within and across teams are essential to avoid mistakes and errors. The ways in which responsibility for services is allocated to different groups within the equipment management system vary widely, even within what are considered national health systems. Traditionally in the United Kingdom, clinical engineering services were responsible for medical equipment, estates and works departments for the physical infrastructure and information technology (IT) managed information and computing networks and services. In reality, life is more complicated. For example, medical gases are classed as drugs and fall under the supervision of a pharmacist but are transported in cylinders by porters or through pipe networks overseen by estates departments, and delivered via regulators supported by clinical engineering. Outside contractors or manufacturers who are responsible for the maintenance of medical equipment are likely to liaise not only with users but also with clinical engineering, and sometimes estates and IT, to ensure facilities are in place. Modern devices increasingly use computer networks, and it is not unusual to run into interfacing problems where clinical engineering has to facilitate communication and clarify responsibilities for solving an equipment problem between a manufacturer and the IT department.

The responsibility for looking after and solving problems associated with unglamorous equipment may be overlooked in the overall equipment management system or be subject to disputes between groups claiming or denying responsibility for them. For example, patient and equipment trolleys are prone to mechanical instability due to incorrect loading or develop simple mechanical faults, yet are rarely serviced unless supporting more complex equipment; portable electrical socket outlets are discouraged but often in frequent use and subject to little care or inspection; and patient monitor wall mountings can cause serious patient harm if they give way but are not always checked during equipment servicing. Simply writing a duty of care into the medical device policy will not guarantee everything is covered and such items require clear and positive assignment of responsibilities regarding who will inspect them, with action if problems are found. Potential risks are easily missed or passed over, and incidents are caused by simple mechanical or electrical failures that persist because nobody is directly responsible for sorting them out. Incident investigators should be on the lookout for issues such as these, and recommendations to address them will emphasise the need to negotiate who is responsible for what and is likely to require meetings and discussions within and between groups, and even arbitration at a senior level.

Incident investigations often uncover poor communication as a root or contributory cause. Sometimes this is due to individual inadequacy or oversight, but it can also be due to inherent difficulties in the whole way the system works. Systems should be designed to facilitate information flow and limit the extent to which poor communication can disrupt the outcome. Standardising the way information presented by using forms and tick boxes, rather than free text, can discipline staff to focus on key information and improve their ability to read and assimilate its meaning.

### 10.6.5.6 Unexpected Events

Unexpected events cause distraction and can interrupt procedures. Yet *unexpected* need not mean *unforeseen*. Even where events are totally outside the operator's control, such as fire alarms or power failures, their impact can be reduced by prior planning. Asking the question, 'What can go wrong?' when devising procedures provides a good basis for drawing up contingency plans, to either adapt the procedure or manage an orderly shutdown when backup resources are not available. Every system should address what to do if things go wrong, at least for possible interruptions such as equipment or power failure, and that includes imagining the unexpected. Particular areas of potential risk include the introduction of new procedures and the use of equipment off label. Examples of such occurrences [16] emphasise the need for caution, lateral thinking and a comprehensive risk assessment before undertaking a new procedure or varying an existing one. The unexpected should be taken into consideration when planning protocols or procedures.

## 10.7 Devising Control Measures

Control measures aim to prevent or reduce the undesirable impact of unsafe situations caused by human error, equipment breakdown or system failure. Each procedure may have multiple control measures, and systems fail when a number of these are breached, leaving the system open to an incident occurring. The *Swiss cheese* model [7] compares risk control measures to layers of protection, each one having potential flaws or workarounds (holes) via which errors occur. Putting these layers together like stacked slabs of Swiss cheese means, most of the time, that holes or flaws in each layer are hidden by adjacent control measures. On occasion, when multiple measures fail, the holes line up right through the stack, the defence system is breached and a cumulative error occurs. This model illustrates the truth that incidents occur when multiple flaws exist simultaneously. An incident cannot occur if one or more safety features are placed. However, staff facing work and time pressures may relax inhibiting safety systems one by one, until the stage is reached where failure of just one last safety precaution can result in an incident. There is an underlying, if flawed, logic in doing so as omitting a single safety step is unlikely to result in an incident – so users learn that occasional laxity does not lead to disaster. Constant vigilance and reinforcement is the only way to overcome the human tendency to minimise effort.

Control measures must be considered when implementing a new procedure and are likely to be modified as a result of incident investigations. Any modifications should address any system failures identified. Control measures can work both by forbidding certain actions and by making other actions compulsory. Failure mode analysis identifies the ways in which a system can fail, the possible consequences and how failures can be directed into safe modes. For example, if a building's power supply fails during a fire, any external doors normally held locked by an electromagnet will be released, avoiding the possibility of trapping people in the building. This is a failsafe system. Control measures can aim at total prevention by, for example, limiting access to systems by untrained personnel or the public. They can also protect operators from undue strain or distraction by preventing interruptions and by monitoring and limiting workloads or hours worked. Action-limiting systems aim to reduce the potential for and effects of human error. Whilst unsafe actions can be made impossible only by discontinuing the use of the device or procedure, limiting measures can make unsafe actions more difficult by the following:

1. Restricting access to certain device functions or software, via policies, procedures, keys or passwords, so only authorised and trained users are allowed to carry out required actions and deal with foreseeable contingencies. In addition, the most sensitive tasks may require the presence of senior and experienced staff who are more likely to identify developing problems and who have the authority and experience to respond appropriately to developing problems.

2. Blocking certain combinations of device control settings that could be hazardous.

3. Checking actions in advance, by getting a system to query the operator before continuing. An item of equipment may display a message like, 'Doing this will … are you sure you want to …?' when it receives a command that might lead to an unsafe condition. Safe use can be encouraged by the following:

   a. Removing ambiguity, for example, making sure device controls and instructions are clear.

b.  Providing standard operating procedures and monitoring if they are being followed.

c.  Requiring pre-use checks to verify equipment is working and check for obvious safety breaches. These also remind users of important equipment functions and controls.

Another control measure is to make sure all necessary resources are in place before allowing procedures to go ahead. These might include checklists of equipment, consumables and spares; pre-use checks for user maintenance; and visual inspection to ensure equipment is in good condition and calibration is in date. Checks would also include the availability of backup equipment, alternative clinical procedures and emergency resuscitation kit. For example, if a life-critical ventilator were to fail in an intensive care unit, a spare unit should be available together with equipment to ventilate the patient by hand until a replacement ventilator can be set up. Complex tasks are inherently more dangerous. Relying on an operator to memorise and correctly reproduce a long and complex operating sequence is asking for trouble, and a control measure in this case is to break tasks down into smaller, simpler components and provide a checklist to track overall progress.

## 10.8  Outputs from an Incident Investigation

An open incident investigation report will be published, archived and made available to staff and managers and possibly to external bodies such as notified bodies or manufacturers. Where an external body has carried out related investigations, its report will need to be incorporated into any internal report and its conclusions. If an investigation is likely to take a long time, an interim report may be produced to provide short-term recommendations or reassurance prior to the final report. Any recommendations for policy change will need to be incorporated into the organisation's redrafted policies and procedures, after consultation.

If an enquiry identifies obsolete, deteriorating or otherwise dysfunctional equipment as being a significant risk, the organisation will consider whether and how to replace it and will factor this requirement into its resource planning and allocation. Urgent replacement may require emergency funding or revision of the equipment replacement programme. Where a change in the nature or frequency of maintenance, inspection or quality assurance protocols is indicated, urgent discussions will be needed between clinical engineering and the manufacturer, third party service agent or multiservice vendor as appropriate, to ensure the new approach is implemented.

## 10.9 Monitoring

Monitoring should be set up once any changes have been introduced, to check whether revised processes are being followed and achieve their original aims, with audit of user compliance. Process development is iterative, and reviewing the effect of initial changes often uncovers further improvements that can be made. Monitoring and follow-up therefore need to continue for long enough to embed new control measures and systems firmly. Staff should be encouraged to report further similar incidents or near misses related to the original incident. Subsequently, the organisation should query its incident reporting system regularly to identify any further problems in the relevant area.

Monitoring may uncover serious failings in recommended control measures and other system changes, including poor user compliance. If this is the case, the organisation should convene a panel to review progress, look at experience in other organisations and suggest further actions to reduce the frequency or severity of incidents and near misses. It may even be necessary to devise an action plan to trigger withdrawal of a service, immediate replacement of equipment or other urgent action if monitoring shows that this is necessary.

## 10.10 Summary

This chapter highlights the important role incident investigation plays in improving patient safety and reducing organisational risk. It sets out issues to look for when embarking on an incident enquiry and techniques to help with its careful and structured handling. Human, equipment and systems factors are highlighted alongside the need to take a systems approach. Finally, it presents ideas for devising risk control measures and stresses the role of continued monitoring in making the organisation's processes safer and more effective.

## References

1. World Alliance for Patient Safety. *WHO Draft Guidelines for Adverse Event Reporting and Learning Systems*. WHO, Geneva, Switzerland, 2005.
2. Radiation Oncology Safety Information System (ROSIS). http://www.rosis.info/ (accessed on September 06, 2013).

3. Jacobson, B. and Murray, A. *Medical Devices—Use and Safety*. Churchill Livingstone, New York, 2007.
4. MHRA. *DB 2011(01)—Reporting Adverse Incidents and Disseminating Medical Device Alerts*. Crown Copyright, London, U.K., 2011.
5. Bronzino, J. (ed.). *The Biomedical Engineering Handbook*, 2nd edn. CRC Press, Boca Raton, FL, 2000.
6. Institute of Physics and Engineering in Medicine. *Report 95: Risk Management and Its Application to Medical Device Management*. IPEM, York, England, 2008.
7. Reason, J. *Human Error*. Cambridge University Press, Cambridge, U.K., 1990.
8. Edwards, E. The importance of utilizing human factors engineering in developing biomedical innovation. Intelliject, LLC. http://nciia.org/conf08/assets/pub/edwards.pdf (accessed on September 06, 2013).
9. UK National Reporting and Learning System. *Design for Patient Safety: A Guide to the Design of Electronic Infusion Devices*. NRLS, London, U.K., 2010.
10. Amoore, J. and Ingram, P. Learning from adverse incidents involving medical devices. *BMJ*, 325, 272, 2002.
11. ECRI. Top ten health technology hazards for 2013. https://www.ecri.org/Documents/Secure/Health_Devices_Top_10_Hazards_2013.pdf (accessed on September 06, 2013).
12. Hillman, B. and Goldsmith, J. The uncritical use of high-tech medical imaging. *N. Engl. J. Med.*, 363, 4–6, 2010.
13. Goldacre, B. *Bad Science*. Harper Perennial, New York, 2009.
14. Bagnall, A-M., Wilby, J., Glanville, J. and Sowden, A. *Scoping Review of Sabotage and/or Tampering in the NHS*. Centre for Reviews & Dissemination (CRD), University of York, York, England, 2004.
15. Henderson, J., Willson, K., Jago, J. and Whittingham, T. A survey of the acoustic outputs of diagnostic ultrasound equipment in current clinical use. *Ultrasound Med. Biol.*, 21, 699–705, 1995.
16. MHRA. *MDA/2010/001—Medical Devices in General and Non-Medical Products*. Crown Copyright, London, U.K., 2010.

# 11

## Supporting Research and Development

### 11.1 Introduction

Medical device innovation is often thought of as involving the modification, development or construction of medical devices. It also includes their introduction or application in a new role as part of research and device evaluation, or to support a new or novel clinical technique. Clinical engineers may create, or participate in developing, new medical devices but in a hospital are most likely to be responsible for managing the underlying policies and processes that make sure devices produced by others are used safely and effectively. Their involvement may comprise anything from carrying out acceptance procedures on innovative devices to undertaking in-depth effectiveness evaluations or managing an entire device-orientated research project.

We use the term *innovation* to cover all those situations where an organisation is aiming to introduce devices into the clinical environment that either cannot be purchased off the shelf or which are being used in ways not originally intended or approved. A clinical engineer has three principal functions with regard to medical device innovation:

- Accepting (or rejecting) innovative devices for use within the organisation, after assessing whether regulatory, governance, safety and risk management requirements have been met sufficiently.

- Assisting innovators, by providing expertise, technological advice and a knowledge of regulations and standards to help bring a device or system to the point where it will comply with relevant technical standards and meet criteria for clinical acceptance. Assistance can include acting as a link between an external designer and clinical staff to clarify required specifications and may extend to constructing or modifying items of equipment.

- Initiating or playing an active part in the development and construction of novel devices or systems, and in finding new uses for existing and innovative technologies.

Issues a clinical engineer is likely to encounter when dealing with novel devices include the following:

- Medical devices making measurements or providing interventions in line with their intended purpose but applied to a research use.
- Off-label use of a device licensed for clinical application but being applied for another purpose.
- Introduction of non-medical-grade items, for clinical or research use, legally placed on the market but not registered for medical use.
- In-house modifications or new equipment built or developed by the organisation, including software, to meet local clinical needs or solve a particular research problem. This includes novel items custom built for individual patients.
- Items not approved as medical devices in their own country but sold legally elsewhere.
- Items used, made, modified or assembled in-house to facilitate clinical research which are not medical devices in their own right.
- Items intended for clinical investigation by manufacturers prior to placing on the market.

In this chapter, we discuss the management of all these device categories. First, we consider how an innovator in a health organisation can legitimately undertake projects by gaining organisational approval, setting up appropriate control and risk management processes, obtaining external approval by ethics and regulatory bodies and complying with law and regulation. The clinical engineer must be aware of all these processes in order to make sure that they avoid unwittingly supporting projects of dubious value to the organisation or those which inadvertently contravene legal or governance requirements. Secondly, we show how clinical engineering services can contribute to the management of equipment innovation, by developing appropriate policies and procedures dealing with non–standard devices, and by offering technical assistance to innovators to optimise performance and minimise risk in novel devices or applications. Thirdly, outline how organisations might go about constructing their own innovative devices, should they need to do so.

## 11.2  Legitimising and Managing Research Projects

### 11.2.1  Internal Governance

All innovative medical device use requires organisational scrutiny in one way or another. Even minor changes to equipment or routine procedures should have some form of approval for corporate governance purposes, with a record of who made the decision and why. If equipment is introduced or altered by people without sufficient expertise to understand the consequences, this can lead to injury to the patient and potentially prosecution for the clinician (see Chapter 10). Healthcare organisations should have policies governing the introduction of new devices and procedures into routine clinical use and for bringing research equipment into the clinical environment. Where innovation occurs in a research context, each project will require specific approval through research governance structures and processes.

A formally constituted group such as a local risk committee will need to review and approve the clinical use of new and modified medical devices within an organisation. Alongside this group a formal research committee considers and approves the introduction of new research trials, and in large organisations this group is likely to be supported by a dedicated research team which provides supporting advice. These committees may approve or reject the introduction of new procedures, devices or research on several grounds. Questions they will ask include the following: Does it fit with the aims and objectives of the organisation? Is there an viable business case, with appropriate financial operational and project management planning? Is there adequate funding to reimburse any additional costs incurred by routine clinical services and have associated risks and liabilities been addressed? Is it safe? Is it ethical?

The clinical engineering service may play a part in introducing equipment for both innovative clinical services and research programmes. A clinical engineer who develops new equipment and techniques to be tried on human beings will need to follow the same research principles and processes as any clinical investigator, even though the project may involve only equipment tests on healthy volunteers. Also clinical staff acting as investigators may be unsure of how to comply with medical device law under varying circumstances, and a clinical engineer with good background knowledge in this area can provide advice to help avoid taking undue risks or exercising excessive caution.

## 11.2.2 Ethical Considerations

Two things are absolutely necessary where devices are used for research on human subjects: the study must have ethics approval, and individual subjects must give signed, informed consent. Consent indicates the subject understands the background to the research, that details of the procedures they will undergo have been explained to them, and that they have received a clear statement of any benefits and risks involved.

Clinical research activity on humans has to be cleared with outside bodies. Ethics bodies in the United Kingdom comprise medical and other scientific experts, together with lay members. The experimental protocol must demonstrate adherence to basic ethical principles [1] and describe in detail what will happen at every stage of the research procedure, for all possible experimental options. Applications should also include proposed patient documentation and consent forms. In the United Kingdom, the National Health Service Integrated Research Application System (NHS IRAS) form is used to process project applications. It covers both ethics approval for human experimentation and approval from the UK regulatory body, the Medicine and Healthcare products Regulatory Agency (MHRA), for the investigational use of devices. Approval is granted for a certain period during which the work must be carried out to the agreed protocols. The MHRA requires detailed technical details of equipment used and evidence it complies with regulations, in much the same way as when a manufacturer places a device on the market. However, research equipment used only inside an organisation may not need external approval, only risk management procedures, and this is discussed in Section 11.3.

Where a clinical procedure obtains samples or data for clinical purposes, any concurrent or future research use of that data must have the patient's consent. Some organisations have obtained generalised ethics approval to use anonymous data such as medical images or physiological recordings for research or teaching purposes, but otherwise ethical body consent is required for each individual study before clinical data are generated or used for research.

The earlier stipulations give a brief overview of what is required. For a detailed description of requirements, the reader should consult the websites of their national research ethics bodies such as the UK National Research Ethics Service [2] and also look at advice and guidance set out by their own healthcare organisations.

### 11.2.3 Confidentiality

Experimental data must be managed according to strict confidentiality criteria, with data anonymised and restricted to authorised persons only. Data must be held securely and be maintained in an agreed storage location for no longer than a specified period. Any use of clinical data for research purposes must be carefully controlled, with anonymisation or coding and encryption of identifiable patient content and clear separation between research and clinical data. A basic principle of confidentiality is that only sufficient personal information is divulged to anybody to enable them to carry out their legitimate work.

### 11.3 How an Organisation Manages Risk Associated with Innovative Equipment

Three main factors determine how innovative devices are managed: whether they are intended for clinical or research use, whether they are used solely within the organisation or in more than one centre, and whether or not they are approved medical devices. A common question that arises with novel equipment is, 'Do we need to get approval from the competent authority?', and the answer depends critically on these three factors.

### 11.3.1 Approved Medical Devices Used for Research Rather than Clinical Applications

Where standard medical devices, such as CE-marked equipment in Europe, are used to make measurements or provide interventions for research rather than clinical purposes, there is no need for approval by a competent authority such as the MHRA if equipment is used for its intended purpose. The organisation's standard approach to acceptance, maintenance, support and training should be followed, as for any item of clinical equipment, with the addition of specialist training in the research application and any additional calibrations or checks needed to ensure confidence in experimental data. Often such equipment is loaned by a company sponsoring the research and in this case liability must be agreed. The NHS has a Master Indemnity Agreement and forms to help ensure responsibilities are defined and appropriate insurance is in place [3].

### 11.3.2 Non-Medical-Grade Items Used as, or in Conjunction with, Medical Devices

Non-medical equipment approved as suitable for being placed on the general market, such as computers or printers, may be attached to medical device systems for research (or clinical) use. Equipment such as video monitors is often supplied as part of a medical device system by commercial medical device manufacturers. In this case the complete system will be designed to comply with medical device regulations and will be sold as such, requiring no special management. However, where an organisation attaches non–medical devices to a medical system in ways not specifically planned for by the manufacturer, it will need to identify and address any associated risks. If mixed medical and non–medical device systems are set up in-house, it becomes the investigator's responsibility to comply with relevant medical device standards. The clinical engineering service can provide advice and perform tests to show if this is the case and may have facilities to make any modifications required. Common adaptations include the use of isolation transformers or opto-isolators for devices whose leakage currents exceed medical limits [4], together with provision of appropriate housings and wiring arrangements to avoid any other devices being connected inadvertently. A risk assessment should be carried out based on the risk assessment standard for medical devices ISO14971 [5] to highlight any other special precautions needed, for example, restricting use to specially trained personnel.

There will be times when equipment required for a new clinical procedure cannot be provided by a manufacturer. One example is using an industrial laser of a specific wavelength in a new medical application because no medically CE-marked version is available. Here the non–medical equipment will be used as a medical device in its own right, and so all physical and clinical risks must be extensively appraised and contained by a combination of physical modifications to meet relevant medical-grade standards and by active risk management of the clinical application.

### 11.3.3  Constructing or Modifying Medical Devices for Use within the Organisation

Items may be made or assembled in-house to facilitate both ethically approved clinical research and novel clinical processes. Their management will depend on whether use is local, within the legal entity that is the organisation, or external. If equipment is used solely for research purposes by the legal entity responsible for it, then it is not intended for placing on the market and medical device regulations do not apply. Similarly, equipment constructed for clinical use solely within the organisation is outside the scope of medical device legislation.

Even though a self-constructed or modified device may not be subject to medical device regulation unless it is placed on the market, associated risks must still be managed carefully. A healthcare organisation has a duty of care towards patients, its staff and the public and must comply with a wide range of legislation (see Chapter 3) including health and safety law. Even where medical device legislation does not apply, complying with its essential requirements by following appropriate construction and risk management standards will help protect against civil litigation, minimise risk and ensure a safe and effective product. An organisation that has built, modified or otherwise devised equipment is advised to at least

- Perform a risk assessment of the equipment and its intended mode of use
- Keep a technical file with all relevant specifications and data, including design data, component compliance and traceability and information on the manufacturing process
- Perform tests or obtain certification to show compliance with relevant regulations
- Provide technical manuals
- Provide user instructions and training
- Limit use of the device to its intended research function and protocols
- Restrict access to authorised staff
- Label equipment as *for research only*

The organisation should broadly follow the same standards expected from manufacturers placing equipment on the market but can risk manage any minor non-compliances where compliance would be inordinately time consuming, expensive or otherwise unreasonable in very early prototypes. For example, if an intricate sub-assembly was required that had an integral indicator light of an inappropriate colour, this could be appropriately labelled and warnings placed in the instructions rather than risk damage to the unit by changing the indicator element. There is a fundamental difference between managing the risks posed by a nonconforming research device, used in a controlled environment by a limited pool of experienced research staff intimately involved in its conception, and those created by releasing a device into general use where its clinical application can no longer be controlled by the designer. In the latter case, nothing less than total conformity with regulation is acceptable.

### 11.3.4 Clinical Investigations of New Devices Prior to Placing on the Market

Where a device that uses a novel clinical technique is developed with the intention of placing it on the market by a commercial manufacturer, it must have undergone a clinical investigation at some stage, to prove that the manufacturer's claims for its clinical performance are justified before it can be approved for sale. The United States Food and Drug Administration (FOA) will, under Section 501(k) of premarket approval, judge and accept the adequacy of prior experience and historical experimental evidence [6], and the UK MHRA has an equivalent process [7]. Clinical device investigations often take place in hospitals, where the manufacturer provides a prototype device, instructions and initial training and supports clinical and scientific staff using it under controlled experimental conditions. Every facility in which a trial takes place must make sure responsibilities and liabilities of all parties for equipment are clearly defined and covered during the lifetime of the investigation, including indemnity for loan equipment. Ethical and any required competent authority approval are also essential. Any medical device on trial should be accompanied by the same level of documentation as any medical device within a healthcare organisation, with user instructions and training provided before use, acceptance procedures followed, and a mechanism set up to report any adverse incidents associated with the trial.

### 11.3.5 Use Outside the Region of Approval

In most countries with medical device regulatory systems, it is illegal to place an item on the market which is not an approved medical device for that country, even if approved for medical use in another region. Similar considerations apply to research devices and those awaiting approval, and this can be a serious barrier to international collaboration on research projects involving new devices. However, if a device is required for use on a specific patient in a single organisation, then it may be possible to use it legally on humanitarian grounds. Some regulatory regimes provide a process for approving such use; in the United Kingdom, application can be made to the MHRA on the grounds of *protection of health* for a medical practitioner to use a specified individual item of non-CE-marked equipment on a named patient, where no suitable approved device is available [8]. Permission is granted only where there is strong justification of significant benefit, such as reduced morbidity or mortality compared with other available treatments, and evidence of a thorough risk assessment.

### 11.3.6 Equipment for Off-Label Clinical Use

Approved medical devices are used in the overwhelming majority of routine clinical procedures. Sometimes, however, a new or improved clinical procedure requires novel equipment or data processing techniques. There may also be circumstances under which custom items may be built for individual patients. Meeting either requirement could necessitate modifying or building equipment and signal acquisition hardware, designing custom software to be used in-house or incorporating non–medical items (often computers) in medical device systems.

An alternative solution is to take a device licensed for one clinical purpose and apply it to another. This is *off-label* use. Whilst not necessarily illegal, this practice is likely to transfer liability for any injury from the supplier to the user but does not require a formal clinical investigation. Any risks associated with off-label use should be identified in a formal risk assessment by the user and discussed with local risk managers and clinicians, to judge whether the benefits justify the action. This is harder to do for research applications than it is for clinical procedures where there is no licensed alternative, such as using some adult devices with paediatric patients.

## 11.4 Practical Aspects of Getting Novel Medical Equipment into Use

The clinical engineering service should support the development of policies and procedures within their organisation which manage the risks associated with novel medical equipment without stifling innovation. Risks should be constrained by formal risk management processes, based on medical device regulatory requirements for the relevant class of equipment. For governance and operational reasons, clinical use of novel medical devices should be approved by a multidisciplinary body such as the organisation's Risk Management Committee. This body should review clinical, technical and organisational risks and benefits before coming to a decision. The clinical engineering service can support this process by evaluating and approving devices on technical grounds. In the simplest cases, this will involve going through standard acceptance procedures and safety checks plus ensuring there is sufficient evidence about the device to meet essential legal, technical and safety requirements and allow clinical benefits and risks to be balanced against each other. Evidence may be provided by the manufacturer alone or through joint working with clinical engineering. In more complex cases, extensive investigation may be required to provide assurance that a device complies with technical standards, particularly for novel devices when appropriate standards may not yet have been developed.

When effective policies and procedures are set up, organisations rather than individuals are the responsible innovators and cover the liability of their employees. For this to work with novel equipment, an organisation must make its staff aware of what they need to do. However, these processes can be seen as annoying sources of delay by researchers and clinicians. Clinical engineering not only should help to streamline the organisation's processes but should also forestall any innocent or contrived attempts to get round them by proactively seeking out any novel equipment or procedures that may have bypassed current processes, or that were introduced without being fully assessed against prevailing regulations.

Acquisition procedures for new medical equipment should also help to identify novel equipment. Possible sources of information on equipment that might have bypassed normal routes include looking at equipment lists for departments where innovation has historically occurred, to assess whether older equipment was self-constructed or has been significantly modified, and visiting areas of the hospital where patients may undergo innovative procedures or take part in university research trials to look at their equipment, including any research, university or independently occupied areas. Cooperation and an open climate of trust can be fostered by building relationships between clinical engineering, the local ethics and research and development offices, innovation and commercial development officers and academic contacts including active research investigators known to be interested in innovative technologies. Valuable information may be gained from estates and facilities and information technology departments, who are likely to be involved with projects introducing non–medical devices into clinical areas. The clinical engineering service should assess all equipment identified via these routes to make sure it is already managed by an appropriate department in an effective way, with maintenance and further development support available from either the original supplier or a qualified successor or from clinical engineering. The supplier may be a department within the organisation, such as medical physics or clinical engineering, or an external manufacturer that is not part of the same legal entity, such as a university or medical device company.

At some point every item of medical equipment in clinical use has to be accepted into the organisation as being fully functional, fit for use on patients and compliant with safety and other regulations. The clinical engineering service usually acts as the organisation's acceptance gateway, although specialist scientists and technicians in areas such as imaging and diagnostic laboratories may also provide this role locally. A typical acceptance process involves a number of tests and inspections which are usually recommended by the manufacturer but may be varied on the advice of a clinical engineer or independent technical adviser. For example, a researcher may ask a manufacturer to deliver a novel system measuring to a higher level of accuracy than normally specified, or one has additional features that are still under development. The user will need extra tests carried out to verify that performance meets the requested standard and is well advised to take independent advice on how to set these up, particularly as ultimate responsibility for acceptance testing rests with the receiving organisation. Any use of external experts or reliance on manufacturer data should be supervised by clinical engineering or other specialist staff, including checking that advisors are appropriately qualified and experienced to carry out work on novel systems. Local policy and practice may set out other duties as part of acceptance, including training and preparing users for the introduction of new equipment and setting it up for first use, and these should also be followed for novel clinical equipment. Finally, the organisation must record acceptance of novel items on its device database and ensure documentation such as user manuals and safety instructions are delivered to end users, even where these have to be specially written to reference novel features. Records of satisfactory acceptance and user training must be kept throughout and beyond the equipment's lifetime, to counter any claims of negligence or liability for subsequent failure.

Where a healthcare organisation acts as a manufacturer, its clinical engineering service is generally responsible for devising acceptance procedures. This applies both to novel medical devices being used solely within an organisation, and also to items not subject to normal quality assurance and regulatory risk controls such as CE marking, even if design and construction is subcontracted to another body. For example, if an organisation wants to carry out research using a medical device which is widely sold in one market but not approved for sale as a medical device in its own country, it will have to assure itself and be able to demonstrate that risks are minimised for the intended use. When planning acceptance procedures for non–approved equipment, a clinical engineering service can follow processes similar to those adopted by regulatory bodies. Although precise requirements vary from country to country, the extent of documentation and proof of compliance to national standards demanded from manufacturers generally increases as medical devices become more complex and present greater risks in use [9]. For example, reference documentation for a medical device classified as Class I under the medical devices regulations in Europe, such as a novel design of operating retractor, might consist of a simple technical file containing annotated drawings and risk assessments. Acceptance testing of a novel device with this classification might then involve running through the risk assessment to check if it applies to the intended use and keeping records of training and initial device use, all signed and dated by qualified and experienced staff. For medical devices of Class II and above, the supplier will provide a lot more information, including technical data on individual components and assemblies, detailed risk assessments and analysis of potential failures to international standards such as ISO 14971, a risk management plan, and evidence of compliance with essential require-ments of national medical device legislation regarding electrical safety and legal requirements for electromagnetic compatibility. External testing or assessment by a competent body or authorised testing organisation may be needed, depending on the class and risk analysis associated with the device, and comprehensive records of these tests will need to be kept.

It is advisable therefore to involve the clinical engineering service as early as possible in planning acceptance testing, particularly for novel situations or where there is any departure from the equipment's standard configuration or intended use. The clinical engineering service should be able to under-take, assist with or advise on any part of the acceptance process, including compiling documentation to evidence compliance and performing, or com-missioning external experts, to carry out appropriate testing. The acceptance process should be set up wherever possible as a mutual arrangement with the supplier, to make best use of available skills and minimise duplication of work. The clinical engineering service will depend upon the expertise, resources and skills available to it and, particularly in a large organisation with a lot of specialist activity, will need to be able to find adequately skilled individuals from elsewhere to assess any complex equipment being pur-chased or developed. The clinical engineering service may also need to mod-ify equipment or provide additional accessories such as earthed enclosures or isolation transformers, in order to ensure compliance to safety standards in a particular application. The degree to which such work is undertaken in-house will depend on the professional skills and knowledge available and is likely to vary as individuals gain experience or leave the service. As an abso-lute minimum, the service needs to ensure the organisation understands its responsibilities for safety and the potential liabilities of not carrying out adequate acceptance testing.

## 11.5 In-House Construction of Novel Devices

Medical physics and clinical engineering services have a long and distin-guished history of leading medical device innovation and application, particularly from within academically connected hospitals. Such technical developments in the United Kingdom include medical ultrasound, elec-trical impedance tomography and magnetic resonance imaging. With the continued evolution of regulation and technical standards, and the increasing cost of advanced research and development, most medical device manufacture and technological innovation now take place in the private sector. At the same time, academic and public institutions have become more aware of the value of their intellectual property, leading to a greater formalisation of the relationship between manufacturers, researchers and the clinicians with whom they collaborate and whose work they might use. An increasing number of start-up companies are being set up by universities to exploit and develop their ideas. In this climate, technological development within healthcare institutions is being channelled towards the translation of new discoveries into clinical service, often funded explicitly through research and technology support grants.

In-house manufacture of devices is undertaken by the relatively small proportion of clinical engineering departments having the processes, procedures, expertise and quality systems capable of producing prototypes or finished devices. It tends to take place in centres where service and academic demands sustain a critical mass of expertise and resources, often associated with substantial amounts of in-house maintenance and support for research and teaching. A greater number of clinical engineering services have sufficient expertise to modify medical devices, and services increasingly collaborate with specialist suppliers, manufacturers and university departments to facilitate the use of existing technology rather than attempting to drive technological innovation. The ability to produce devices in-house depends on having sufficient electronic, computing and mechanical workshop resources, expertise in medical device and research governance and local clinical and organisational support for innovation. Staff must have the skills, knowledge, imagination and stamina to devise novel solutions and sustain the process of development. Resources required for prototype development include access to specialised design, construction and test facilities. However, equipment can still be developed by departments without sophisticated or specialised engineering facilities if they can outsource engineering construction and even detailed technical design, either by taking on subcontractors or through cooperating with manufacturers. Under these circumstances, an in-house service can add value by working with clinical staff to ensure any device or software is specified and designed correctly, risk assessments are carried out and manufacturing quality and output is monitored.

Where finance is available but in-house expertise and resources to develop hardware or software is lacking, it may be possible to outsource all or part of the development process. For example, the production of prototype printed circuit boards to a high quality can easily be performed by external contractors at a reasonable price, eliminating the need to maintain these facilities in-house. Contractors can also perform all other stages of electronic production, even down to creating a finished article from a circuit diagram. Putting fabrication of mechanical assemblies to external contractors requires a greater level of expertise, due to the interaction between design and suitable material properties and tolerances. It also needs contractors who can deliver what is required, and the in-house service will need to be able to verify whether a final product meets specification. In principle, the whole design and production of a prototype could be handed over to a contractor, and it is up to the developer to decide how much to outsource, balancing cost, in-house resources and how well developed ideas are for practical implementation. The more confident an innovator is about the design of their invention, the easier and cheaper it is to outsource. In the proof of concept stage, where ideas are developing quickly and mistakes are being made, significant design changes can take place faster than an external contractor can construct or update a prototype. Managing this constant change can be very expensive and time consuming, and outsourcing requires significant skill and attention to detail to be able to specify a suitably flexible prototype. A good compromise is to pilot speculative projects in-house using relatively crude construction techniques. As the desired design reaches maturity, including detailed design for safety and conformity with regulations, it can be outsourced confidently in the knowledge that the expense and effort required to prepare specifications will not be wasted. It will be interesting to see how technological developments in rapid prototyping and 3D printers affect this balancing act.

## 11.6 Creating a Novel Device

There are four stages to developing and producing a novel medical device and attaining regulatory or governance compliance for use within the organisation. These stages are as follows.

### 11.6.1 Evaluating Whether to Buy, Modify or Build

These options are of increasing cost and complexity. If an existing device does what is needed but is not approved for the desired application, the simplest option is to see whether it can be used off label. Another possibility is to modify an existing device to meet the in-house specification, in a way that can be either approved by the manufacturer or adequately risk assured in-house. This might involve linking together pre-existing devices, whether medical or not, into a system which can be assured for safety. The most complex and expensive option is to construct a completely new device, whether in-house or in partnership with an external organisation. This route can include commissioning a manufacturer or external research consultancy to produce a suitable design and construct an end product. Relevant questions to ask when choosing between these options include the following: Are sufficient resources available to develop a new device or construct it in-house? Should part or all of the design or construction be outsourced? Is a manufacturer interested in making or modifying an existing device? In-house manufacture of a clinical device should only be pursued when there is no other way of achieving the desired result, whether on the grounds of novelty, achieving a desired function that is not available elsewhere or fulfilling a clinical need for a product that is vital but not commercially viable. It should never be used solely as a means to save money by avoiding buying something already commercially available. It is very unlikely that one-off items or limited runs can be made at a true cost below that of a manufacturer's production model. If any device exists that can totally fulfil clinical requirements or be easily adapted to do so in a way which conforms with local medical device regulations, then in-house production will not be justified and may expose the organisation to indefensible liability.

### 11.6.2  Establish Resources and Funding

A business plan should be developed and approved, along with resources to implement it. Money, people and time will be needed, and projects that proceed through several stages may need multiple iterations and repeated review during their lifetime. Resources may be provided in-house or come from a partner, manufacturer or grant-funding body or even a venture capital company for a promising idea which has shown proof of principle. Small start-up funds may be available within the organisation to support or pilot projects costing a few thousand pounds, and suitable funds may be directed specifically towards the achievement of specific clinical or technical goals in order to encourage innovation in various areas of practice.

### 11.6.3  Designing for Conformity

A final device design must provide the specified clinical functions. It should also meet all quality control, risk management and other requirements for safe operation to a standard equal to that needed for regulatory compliance. Although initial and crude proof of concept prototypes may be produced without these wider considerations, the earlier they are included in the design process, the lower the likelihood of having to carry out an expensive or time-consuming redesign. Likewise checking sub-assemblies and components at an early stage for compliance with regulations and validating their construction and performance can save expensive remedial works later on. For later stages of the development process, it is important to check that components are available in the restriction of hazardous substances (RoHS) compliant form [10] and not to design on the basis of older components conveniently to hand, as might be done when building early prototypes.

### 11.6.4 Production, Validation and Traceability of Components and Suppliers

If an organisation intend to proceed to build and market medical devices, the design process will need to be incorporated within the same quality management system associated with production. Traceability of design, even at an early stage, can be facilitated by keeping sketches and calculations and maintaining notes of meetings with customers (the design review process). For a design, manufacturers must provide objective evidence of how it meets the essential requirements for regulation. This technical information should be held within a technical file or design dossier. For medical devices manufactured under a regulatory framework, a documented quality system must be in place to ensure devices coming out of production continue to comply with these essential requirements and are consistent with information contained in the technical design file. However, validation of conformity is possible without a formal quality management system if each device is tested and examined individually. For placing on the market, or for a clinical investigation, the manufacturer should obtain documentation demonstrating components, sub-assemblies and manufacturing processes conform to essential requirements. The specific conformity requirements covering custom-made devices and devices intended for clinical investigations in the EU are set out in Article 11 of the Medical Devices Directive [11]. Tests must be performed to show a device completed to the design and manufacturing process specified then complies with relevant standards, including those for electrical safety and electromagnetic compatibility, with results verified by tests carried out by a notified body where required.

---

## 11.7 Placing on the Market

When placing a medical device on the market, manufacturers, including an in-house service, must typically ensure that they

- Comply with the essential requirements relevant to their device
- Demonstrate the design delivers what is claimed
- Carry out a risk assessment
- Demonstrate clinical evidence of the effectiveness of the device
- Classify the device and, for anything other than Class I, obtain a certificate of conformity from a notified body
- Implement a procedure for post–market surveillance
- Complete a declaration of conformity
- Maintain a file of technical information about the product
- Put a CE mark on the product

The associated technical file must contain at least a declaration of conformity and classification, the name and address of the manufacturer and a product description, specification and details of verification. The declaration of conformity contains a statement by the manufacturer that the device complies with the essential requirements of the directive and also states the standards complied with and the device classification. Certificates of conformity are also included.

The *manufacturer* is not necessarily the person who physically constructs the device but is rather the person responsible for placing the device on the market or into a clinical investigation. This person ensures and states conformity with the essential requirements and is responsible for post–market surveillance and any product recalls. The *product description* may include device classification, intended use, whether the device is intended for single use only and a statement of any restrictions on its use. It will contain the modes and principles of operation. The *product specification* is likely to include detailed technical and performance specifications, including a programme for assessing and managing risk during the lifetime of the device. The latter will include an in-depth risk analysis and also a failure mode and effect analysis. Also included will be circuit diagrams, mechanical construction details and drawings together with other relevant documentation such as technical manuals and user instructions, including methods of decontamination. The user manual must contain sufficient information to enable the user to operate the device safely and effectively, including any user maintenance required such as regular cleaning. The service manual supports authorised technical staff in carrying out maintenance, troubleshooting and repair operations. In practice, user instructions coupled with a statement of the user training required may between them provide a good deal of the information required for the product description and specification. *Product verification* contains details of procedures and tests designed to demonstrate conformity with essential requirements and design and performance specifications. It includes information about production quality control and testing, clinical validation or reference to existing clinical data and tests or examinations carried out by an accredited facility to demonstrate conformity with relevant standards.

*Post–market surveillance* includes a means of reporting adverse incidents to the competent authority. Each manufacturer must have a procedure in place to monitor what is happening to their products once sold and to pick up reports of adverse incidents. In the EU a formal vigilance procedure is established by the medical devices regulations through which manufacturers, health professionals and others must report certain problems which arise in the use of medical devices and in clinical trials. These include reporting serious adverse events, injuries or deaths in which a medical device is involved and can in some cases lead to a device recall. Proposals to strengthen surveillance systems in both the United States and the EU have been put forward [12,13].

## 11.8 Summary

Medical device innovation can take many forms, whether the trial of a novel device for the first time in a clinical setting, use of an established device for a new clinical purpose, use of non–medical devices in conjunction with medical devices or developing software to perform a new clinical function. Clinical engineers have many roles in innovation, including accepting novel devices on behalf of the organisation, advising innovators or themselves initiating new developments. In this chapter we have seen how these functions can be carried out to minimise risk but not stifle development and how policies and procedures within an organisation should actively manage innovation. We expanded on medical device regulation, with particular regard to innovation and how devices modified or built in-house can legally be used on human subjects in research. Finally, we looked at how an in-house service might approach the creation of new medical devices and make them available to others.

## References

1. Council for International Organizations of Medical Sciences. *International Ethical Guidelines for Biomedical Research Involving Human Subjects.* CIOMS, Geneva, Switzerland, 2002.
2. UK National Research Ethics Service (NRES). Approval requirements. http://www.nres.nhs.uk/applications/approval-requirements/ (accessed on August 27, 2013).
3. NHS MIA Register for Suppliers. NHS Master Indemnity Agreement Register for Suppliers. http://nhsmia.bipsolutions.com/ (accessed on August 27, 2013).
4. IEC 60601-1. *Medical Electrical Equipment – Part 1: General Requirements for Basic Safety and Essential Performance*, 3rd edn. International Electrotechnical Commission, Geneva, Switzerland, 2005.
5. ISO 14971:2007. *Medical Devices – Application of Risk Management to Medical Devices.* ISO, Geneva, Switzerland, 2007.
6. U.S. Food and Drug Administration. *Device Approvals and Clearances.* http://www.fda.gov/medicaldevices/productsandmedicalprocedures/deviceapprovalsandclearances/default.htm (accessed on August 27, 2013).
7. ISO 14155:2011. *Clinical Investigation of Medical Devices for Human Subjects – Good Clinical Practice.* ISO, Geneva, Switzerland, 2011.
8. U.K. MHRA. *Bulletin 21 – Application for the Exceptional Use of Non-Complying Devices.* MHRA, London, U.K., 2011.
9. Association of British Healthcare Industries (ABHI). Clinical investigations. http://www.abhi.org.uk/key-issues/technical-regulatory/Clinical-Investigations.aspx (accessed on August 27, 2013).
10. European Commission. The restriction of hazardous substances (RoHS) directive 2011/65/EU. *Official J. Eur. Union*, L 174/88, pp. 88–110, 2011.

11. European Commission. The medical devices directive 2007/47/EC. *Official J. Eur. Union*, L 247/21, pp. 21–55, 2007.
12. Center for Devices and Radiological Health. *Strengthening Our National System for Medical Device Postmarket Surveillance – Update and Next Steps*. US Food and Drug Administration, Silver Spring, MD, 2013.
13. MHRA. 2013. The revision of European legislation on medical devices – The response to the public consultation. http://www.mhra.gov.uk/home/groups/comms-ic/documents/publication/con260309.pdf (accessed on August 27, 2013).

# 12

## Disposal

### 12.1 Introduction

All medical equipment will, at some point, reach the end of its useful life. An organisation then has the problem of what to do with it. Disposal may mean physically destroying it for scrap or recycling but can also involve device transfer to another party as working equipment, for refurbishment or for another further use. Old medical equipment should not languish in cupboards, drawers or corridors for years, in case it comes in useful. If not actively maintained, there is a risk unsafe equipment will be used clinically by individuals whose training is out of date and with consumables that are incompatible.

Organisations should proactively manage condemning and disposal of equipment and take responsibility for its final destination, keeping disposal records and informing any necessary authorities of the process. Serious incidents emphasise the possible consequences of getting it wrong. One dramatic example is the Goiânia incident in Brazil in 1987 involving four deaths [1]. A radiotherapy unit was simply abandoned when a clinic closed and its highly radioactive caesium-137 source was removed by scavengers and sold to a scrapyard where it was broken up. Some of the caesium chloride, a glowing blue powder, was taken into the scrapyard owner's house due to curiosity where it was played with by children and used for body painting. Spread unwittingly by visitors, it led to extensive contamination and serious injury to hundreds of people. Such frightening incidents are rare, but as a society, we are increasingly aware of the damage that can be done by the incorrect disposal of toxic materials and complex chemicals that find their way into human and animal food chains or otherwise damage the environment. Particularly prevalent in electronic waste are the toxic metals mercury, cadmium and lead and toxic organic chemicals such as polychlorinated biphenyls (PCBs). If obsolete equipment falls into the hands of irresponsible people, there are dangers not just from toxicity but also from fraud such as attempted resale of substandard equipment or from its use by charlatans or hobbyists.

In this chapter, we consider issues involved in the disposal of medical equipment. Clinical engineering departments are well placed to handle disposal because they have, or can acquire, the technical expertise required to decommission equipment safely and advise which disposal methods are suitable. Clinical engineering departments usually hold and maintain the organisation's equipment records and database, and this must be updated to reflect disposal.

## 12.2  Condemning and Disposal Procedures

An organisation should include a section on disposal in its medical device policy. The decision to condemn equipment and its method of disposal should be governed by the organisation's medical device policy and financial procedures. These may, for example, aim to transfer liability for any subsequent damage or loss attributable to the equipment away from the organisation and ensure its disposal or transfer complies with laws and regulations pertaining to waste materials. It may also seek to realise the maximum residual value of its equipment and support efforts to find alternative users within the organisation. Clearly defined condemning and disposal procedures also help to protect against fraud or allegations of misconduct. The disposal process must be auditable, particularly where items are sold to external users or waste organisations, with the ability to trace transfer of ownership for each individual item or its component parts to the next owner or scrapping service.

The UK government publication DB2006(05) [2] gives an overview of the main issues raised by disposal and describes principles by which hospitals can comply with these requirements. The World Health Organization [3] similarly publishes guidance including how medical equipment and other disposal might be carried out safely in developing countries.

## 12.3  Legislation Relevant to Disposal

Laws govern the disposal of equipment or scrap materials (*waste*) in many countries. They are often multiple and complex, requiring a good understanding in order to ensure compliance. Laws and regulations can relate to the handling, transport and disposal of specific (*hazardous*) substances and groups of equipment and may apply to particular processes or locations. Where redundant equipment is sold or otherwise passed on as working, consumer protection laws can apply which may be enforceable by criminal or civil law or both. Change of ownership of equipment, whether it is working or not, will be subject to agreement under civil law regarding liabilities and responsibilities. An organisation's liability as a supplier is controlled better through a formal contract, under legislation such as the UK Consumer Protection Act [4] (1987), than by reliance on implied obligations under any general legal precedents.

The Resource Conservation and Recovery Act [5] (RCRA) is the principal federal law in the United States governing disposal of solid and hazardous waste. This law and the relevant European Union (EU) directive [6], along with other national regulatory systems, identify and seek to control hazards and are regularly updated as new issues emerge. *Hazardous waste* requiring special treatment is defined by inclusion on a list of such substances or by it having certain hazardous properties defined by regulations. The United States' definition of hazardous waste is based on it possessing certain properties, which can include one or more of the following:

- *Ignitability*: A low flash point or tendency to spontaneous combustion
- *Corrosiveness*: Low or high pH capable of corroding metallic or other containers
- *Reactivity*: Liable to explode and emit toxic fumes, gases or vapours when heated, compressed or mixed with water
- *Toxicity*: Containing greater than permitted levels of toxins

This general definition of hazards allows for the classification of new substances or ones inadvertently omitted from the list, as no list can be comprehensive or remain up to date for long. The relevant EU directive [6] also lists processes defined as giving rise to hazardous materials. Legislation may also prescribe methods to assess concentrations or other properties at which materials become hazardous, such as a flash point or toxicity level, and to determine treatments which can introduce hazards such as the neutralisation of acids by alkalis and vice versa. Regulations are likely to cover the safe production, storage, transport and treatment of hazardous substances, setting out safe procedures, licensing arrangements for persons competent to perform these operations and the documentation required for each process. A register of licensed or recorded producers and disposers is required to ensure traceability of hazardous materials from production to disposal. Similarly, the control and traceability of transport, storage and distribution is usually enacted through a register of licensed carriers and storage facilities, together with documentation such as consignment notes that record the composition and amount of waste stored or transported. In the United Kingdom, a whole suite of legislation applies to the transport of goods by road and rail [7].

Although the legislation introduced earlier is mainly concerned with waste, each country is likely to have specific legislation concerned with the disposal of devices and equipment. Two related directives in the EU are the Waste Electrical and Electronic Equipment (WEEE) Directive [8] and the Restriction of Hazardous Substances (RoHS) Directive [9]. Both apply to electrical and electronic equipment (EEE). Other countries are likely to have similar regulations covering these and other related areas, such as the disposal of batteries [10].

The WEEE Directive covers disposal of consumer EEE and electrical/electronic equipment intended for professional use, including medical devices, with the exception of implanted and infected products. It aims to control the release of hazardous materials into the environment, minimise risks to the environment from such materials, and encourage reuse, recycling and materials recovery. It sets minimum criteria that member states may exceed if they wish. In line with the *polluter pays* principle, it aims to shift payment for the collection and management of waste from general taxpayers to users, via manufacturers. Under WEEE, each manufacturer is responsible for setting up and financing systems whereby medical devices can be returned to them at the end of their life. The onus is on the manufacturer to provide suitable routes to acceptable final disposal. EEE producers in the United Kingdom are required to register with the Environment Agency.

WEEE arrangements for items sold for professional or business use assume a closer and more contractual interaction between producers and users. Equipment is likely to be collected directly or via specialist disposal companies rather than being discarded through a general collection facility. Medical equipment is affected by a further requirement: When an item purchased before August 2005 is discarded and not replaced, the user has an obligation to finance its recycling, recovery and environmentally sound disposal.

The regulations do not prevent any producer or business end user from making their own contractual arrangements which ensure waste EEE is correctly treated and reprocessed. Thus, the purchaser may, through negotiation with the producer, accept the responsibility and future costs of recovery and treatment of EEE they discard in return for a discount on the initial purchase. This is clearly a commercial decision for a healthcare organisation to make. The regulations do not place obligations or requirements on what form any such commercial arrangement will take. In practice, however, it is likely that most healthcare organisations will choose either to pass such equipment back to the manufacturer or to make use of third-party disposal routes. The latter can be cost-effective where equipment or consequent scrap has a significant resale value.

European regulations also aim to prevent certain substances entering the product cycle. The RoHS Directive sets standards throughout the EU restricting the use of certain toxic materials in new equipment, banning placing on the EU market of new EEE containing specified levels of lead, cadmium, chromium, mercury and certain organic compounds. However, many exclusions and exemptions are granted to manufacturers and certain product types where there is no practicable alternative.

These include batteries used in safety or other critical-use applications, so nickel–cadmium cells may be allowed in medical equipment. All new electronic components used in newly built devices and intended for placing on the market must be RoHS compliant. Spare parts containing banned substances may be used to repair equipment manufactured before the implementation date of the original directive (July 2006). The 2011 RoHS Directive covers medical devices but provides exemptions for active implantable devices such as pacemakers and other items where there is no viable alternative material, such as lead radiation shielding in mobile x-ray devices. One implication is that in the future, CE marking as a medical device will also denote compliance with RoHS.

Products built for own use are generally not considered as being placed on the market, and therefore, RoHS does not apply strictly to devices built in-house for an organisation's own use. However, for most projects undertaken in-house, there will be no advantage in not using RoHS-compliant parts and materials as these are now routinely supplied.

## 12.4 Preparing for Disposal

Equipment must be physically cleaned and decontaminated to control infection. If transferred, it should be accompanied by a certificate to confirm this. Confidential information that could identify patients or staff must be erased permanently, whether from data storage components in a device or from associated computers [11]. Patient information which is not encrypted is particularly sensitive. Some data storage systems erase information simply by removing file labels, without wiping the underlying data, allowing it to be accessed subsequently using widely available editing software. Stringent efforts should be made to prevent data becoming vulnerable in this way and may require special techniques such as implementing encryption when equipment is in use or employing specialist erasure software at the end of life. Measures as extreme as replacing hard disks or other storage components may be required to be fully assured of data protection if equipment is to be reused.

Once condemned, equipment awaiting disposal should be kept secure to prevent theft, tampering or data security breaches. It should be labelled to prevent inadvertent clinical use and, if it is in a dangerous condition or intrinsically hazardous, be rendered inoperative by measures such as the removal of mains leads and fuses. Storing bulky equipment in corridors not only breaches these requirements but also blocks or impedes exit routes for fire evacuation. It also encourages general dumping, leading to unsightly clutter and a safety risk to patients, visitors and staff.

## 12.5 Disposal Routes

We now consider the possible disposal routes for condemned equipment, that is, equipment no longer considered suitable or fit for use in the organisation. In practice, the relative benefits and costs of different disposal options must be considered for each item, including any legal requirements, associated costs, complexity and timescales.

### 12.5.1 Return to Manufacturer

Some national or multinational regulations such as WEEE in Europe put a responsibility on manufacturers to make arrangements to retrieve and safely dispose of EEE at the end of its working life. This has the great advantage for equipment owners that they simply need to return equipment to the manufacturer to discharge their legal responsibilities, without having to deal with the complexities of safe disposal. However, some manufacturers may charge for disposal, making this another factor to consider when negotiating equipment purchase. Contractual terms offered by a manufacturer may be more expensive or difficult to adhere to than those which can be negotiated with a third party, and an organisation may achieve better value for money overall by negotiating comprehensive disposal arrangements with specialist company (see the following section). Such arrangements will also be needed for older equipment or when a supplier has gone out of business. Disposal of equipment loaned, leased or operated under a managed service agreement remains the responsibility of the rental or leasing company or service provider, unless the user agrees to take this on.

Another way to dispose of old equipment is to part exchange it for new items. Manufacturers may accept old equipment, including items not of their own manufacture, for a purchase price reduction that may be nominal but which provides good value when the added costs of physical removal and disposal of old equipment are taken into account. This may include upgrade deals when a device is physically replaced by a newer model. In some part exchanges and upgrades, however, the manufacturer may not actually collect equipment but leaves it with service users at their request as a spare, creating a subsequent problem for the organisation to deal with. This underlines the importance of clearly tracking equipment purchasing and replacement in an organisation, to make sure user intentions are clear and that the nominal financial and operational benefits of such arrangements actually do offset the cost of disposal.

### 12.5.2 Using a Specialist Contractor

A number of commercial companies will enter into a contract with organisations to remove and dispose of all redundant equipment, returning a proportion of any profits made from equipment successfully sold or auctioned (either refurbished or sold as seen) and scrapping or otherwise disposing of the rest. These deals are subject to periodic renegotiation because there may be a high proportion of low-value and unwanted items in the redundant equipment and gains from sales do not always cover the overall cost of disposal. However, even if there is a net cost or fluctuation in contract price, the advantage to the organisation is the comprehensive nature of the contract that provides a one stop solution, reduces administrative overheads and transfers liabilities. A specialist contractor must be used for equipment deemed hazardous or subject to specific regulation, such as devices containing radioisotopes.

### 12.5.3 Scrapping

Physical disposal may be carried out by the owning organisation or, as outlined earlier, through disposal routes set up by the manufacturer or a third party. The purpose of scrapping or partial dismantling for an organisation is not so much to recover materials as to render scrapped equipment safe and reduce costs by separating its components into the relevant waste streams. However, some types of medical equipment waste need specialised disposal including electrical components and ones that contain certain metals, some batteries, oil wastes, PCBs, waste coolants, radioactive items and material contaminated by biological materials of human or animal origin. The latter two classes of substance must be directly removed and reprocessed by specialists in these areas. Where applicable, equipment should be biologically decontaminated before disposal or transfer to a third party and be supplied with a certificate of decontamination. If scrapping is carried out in-house, the manufacturer can be asked to identify any hazardous materials contained in the product and to advise on the current recommended methods of disposal. It can also be productive to check if any public or private bodies acting on behalf of governments have set up framework agreements for disposal contracts for specific devices or substances. General advice on disposal is offered by many national regulatory organisations, and this may be particularly important when a manufacturer has ceased trading.

If an external contractor publically states that materials will be disposed of in accordance with certain regulations, then it is not strictly the responsibility of the organisation to check on or enforce this. However, a degree of prudence is necessary, as any unlawful disposal will reflect badly on the commissioning organisation even if it is not responsible in law. As with any contract, periodic checks and audits are desirable to show a duty of care.

### 12.5.4 Use for Spare Parts

*Cannibalisation* involves taking sub-assemblies from condemned or broken equipment to repair equipment still in service. It is discouraged generally by regulators [2] for reasons that include the need to maintain integrity of the equipment record, although this can be addressed through traceability in service records and risk assessment of processes. Experience in practice shows that swapping functioning sub-assemblies in good condition between devices presents little or no additional risk of failure or incident compared to that presented by similarly maintained equipment of a comparable age. The process must, however, be subject to risk control and transparent decisions by competent people.

The reuse of major accessories, for example, ultrasound transducers, which are designed to be plugged into and out of devices, should not pose additional risk as these items are often separately identified in the equipment inventory and can be considered as freely transferable between devices designed to be used with them.

### 12.5.5 Remanufacturing

Remanufacturing occurs when sufficient work is carried out on a medical device or component to replace much of its original material, alter its construction or modify it to be used off label or for a purpose not originally authorised as part of any regulatory approval process. This effectively results in the creation of a new or modified medical device, and all the issues and concerns relevant to this situation will apply. An exception is where remanufacture is carried out according to a manufacturer's instructions, and an example of this is when making a modification approved by the manufacturer to adapt an item for use in a new environment, such as substituting a different voltage mains transformer when selling a used device onto a user in another country. Organisations need to consider their liabilities carefully if they are to engage in this activity, especially if the remanufactured device is likely to end up being owned or operated by a separate independent body. Some third-party maintenance suppliers do this on a large scale.

### 12.5.6 Redeployment for Research and Teaching

Sometimes a research project needs to modify equipment to obtain a level of functionality not available in the original device. This is technically remanufacture, and it happens most commonly with old but still serviceable devices that are surplus to requirements, for two reasons: they are cheaper to obtain and the older technologies they use are more familiar and often easier to modify than more modern, integrated technologies. Manufacturers may also be more willing to divulge technical manufacturing specifications for superseded devices.

It is unlikely that refurbishing or modifying equipment for research use on the health organisation's patients would develop into a placing on the market. However, if the device is modified for sale or supply to other bodies, for example, those involved in a multicentre clinical trial, CE marking and the Medical Devices Directive procedures will need to be followed in the EU and appropriate regulations in other jurisdictions. In an environment where research is being carried out, it can be difficult to determine who has responsibility for a particular medical device, and care should be taken when transferring a device into this environment to ensure that all liabilities are dealt with or transferred and that there is clarity on the eventual route for device disposal. Some types of device may remain in research use out of necessity well beyond their expected or supported life. This use of old equipment is an area seldom addressed by research governance, and organisations should ensure such equipment is risk managed and fit for purpose. Similar principles apply when redirecting medical equipment to teaching. It may be appropriate to use redundant equipment to teach general principles of device operation in an academic context, but careful consideration is needed before training clinical staff on a model no longer in use, as this can lead to confusion rather than clarification where control layouts differ from current model designs.

### 12.5.7 Sale or Donation for Reuse

The EU medical devices regulations [12] cover new or fully refurbished equipment. No EU legislation is devoted specifically to the resale or reuse of medical devices, but there is a substantial body of relevant consumer legislation [4]. Any sold item must be safe and fit for its intended purpose and the sale process must meet certain standards. Hence, it is prudent when selling or transferring a medical device to provide as much information as possible to the new owner regarding the history of the equipment. The information package should include a clear statement that the medical device is being resold or donated, documentation of its decontamination, the original user manual and any training requirements, available service manuals, service histories, usage records and quality assurance test details. Obtaining a disclaimer may limit liability if a device is sold on, but an organisation may still retain some responsibility for contributory negligence, such as poor in-house maintenance work. If a department purchases second-hand equipment, it should take care not to inherit any liabilities due to previous incidents or unpaid purchase costs and should draw up purchase agreements appropriately.

Donating equipment to charities or organisations in developing countries is fraught with potential pitfalls for the receiving organisation. If a supporting infrastructure is not available, including consumables, spares, service support and operating knowledge, then the majority of equipment may never be put successfully into service, let alone continue to work correctly [13]. Guidelines developed by the World Health Organization [14] for equipment donation set out four key principles:

- Healthcare equipment donations should benefit the recipient and be based on a needs assessment and analysis of the environment in which donations will be placed.
- Donations should be given with due respect for the wishes and authority of the recipient and according to a pre-agreed plan.

- There should be no double standard in quality: If an item is unacceptable in the donor country, it is also unacceptable as a donation.
- There should be effective communication between donor and recipient. All donations should respond to an expressed recipient need and should never arrive unannounced.

These guidelines stress the importance of working within a long-term framework and of good relationships and communication between donor and recipient. Ideally there should be personal contact and support to enable successful transfer of donated equipment into clinical service. Good practice is to supply as much information as if the device was being sold for reuse, to ensure that liabilities are carefully defined and that local staff are trained in how to use and maintain the equipment.

Some charitable or other organisations variously refurbish, remanufacture, distribute and even maintain surplus or obsolete equipment, either for second-hand purchase or as donations, for use by healthcare delivery and research organisations in the developed and developing world. Any plans to donate equipment should take account of any support such bodies can provide, but because this is a marginal and unglamorous activity, recent reductions in the availability of free or cheap surplus equipment have caused a number of charities operating in this area to close their operations.

## 12.6 Disposal of Consumables and Batteries

In Europe, consumables only become waste EEE if they are part of another product that falls under the WEEE classification. For example, if a printer is discarded with an ink cartridge inside it, then that cartridge is considered to fall under the WEEE regulations even though it is outside the scope of these regulations if discarded on its own. Used consumables must be disposed of in clinical waste if contaminated with biological material; otherwise they may be managed within other waste streams such as general or domestic waste, as appropriate, or under specified arrangements as for light bulbs or batteries.

Spent batteries must be disposed of via specialist contractors and be separated into different types, including lead–acid, nickel–cadmium, zinc–carbon and lithium-ion. In Europe, the Battery Directive [10] regulates the manufacture and disposal of batteries and bans the use of certain chemicals and metals, sets maximum quantities for others and requires proper waste management including recycling, collection and disposal. Technical problems have limited the scope of implementation because of a lack of suitable alternative materials or appropriate technologies for some applications, with exemptions in place for items such as button cells, which may include mercury, and certain rechargeable batteries used in medical devices on the grounds of critical reliability. The Mercury-Containing and Rechargeable Battery Management Act [15] is a similar law in the United States, banning sale of all mercury-containing batteries except small button cells and requiring labelling for disposal and recycling.

## 12.7 Disposal of Waste from In-House Repair and Manufacturing Activities

This waste stream often includes a wide range of consumables and batteries, to which the previously mentioned considerations apply, together with electronic scrap and scrap metal, plastics and other materials. Electronic scrap may need to be removed by specialist contractors along with other materials and assemblies. In the United Kingdom, the Control of Substances Hazardous to Health (COSHH) Regulations [16] apply to disposal of solvents used in service, repair and manufacture. Proper disposal of all hazardous substances and the scrap contaminated with them is mandatory, and appropriate planning is essential to ensure that this occurs.

## 12.8 Summary

Disposal of medical devices must take place according to good governance and must conform to appropriate regulation. Governance requires traceable decisions to condemn and dispose of equipment. There is a need to reduce the risk of infection by decontaminating disposed equipment and to destroy any personal data associated with it that might breach data protection codes and regulations. Possible disposal routes include return to the manufacturer, sale to a third party, donation, use in research or teaching or scrapping. In every case there is a need to make sure that liability is transferred to the new owner, advantage is taken of any possible cost recovery, and the law is complied with. Many regulations cover the carriage and disposal of waste, and two important European examples are the European WEEE and RoHS Directives.

## References

1. International Atomic Energy Agency. *TECDOC-1009 – Dosimetric and Medical Aspects of the Radiological Accident in Goiânia.* IAEA, Vienna, Austria, 1998.
2. MHRA. *DB2006(05) – Managing Medical Devices.* Reviewed in 2013. Medicines and Healthcare products Regulatory Agency, London, U.K., 2006.

3. WHO. *Medical Device Regulations: Global Overview and Guiding Principles.* The World Health Organisation, Geneva, Switzerland, 2003.
4. UK Parliament. Consumer protection act 1987. http://www.legislation.gov.uk/ukpga/1987/43/contents (accessed on August 27, 2013).
5. US Environmental Protection Agency. Summary of the resource conservation and recovery act (RCRA) 1976. http://www2.epa.gov/laws-regulations/summary-resource-conservation-and-recovery-act (accessed on August 27, 2013).
6. European Commission. Waste framework directive 2008/98/EC, Annex III, *Official J. Eur. Union,* L 312/3, pp. 3–30, 2008.
7. Health and Safety Executive. 2013. Carriage of dangerous goods. http://www.hse.gov.uk/cdg/ (accessed on June 2013).
8. European Commission. Waste Electrical and Electronic Equipment (WEEE) Directive 2012/19/EU. *Official J. Eur. Union,* L 197/38, pp. 38–71, 2012.
9. European Commission. The Restriction of Hazardous Substances (RoHS) Directive 2011/65/EU. *Official J. Eur. Union,* L 174/88, pp. 88–110, 2011.
10. European Commission. Directive 2006/66/EC on batteries and accumulators and waste batteries and accumulators. *Official J. Eur. Union,* L 266/1, pp. 1–14, 2006.
11. UK Parliament. Data protection act 1998. http://www.legislation.gov.uk/ukpga/1998/29/contents (accessed on August 27, 2013).
12. European Commission. The Medical Devices Directive 2007/47/EC. *Official J. Eur. Union,* L 247/21, pp. 21–55, 2007.
13. Howitt, P. et al., Technologies for global health. *Lancet,* 380(9840), 507–535, 2012.
14. World Health Organisation. Guidelines for healthcare equipment donations. WHO/ARA/97.3, 2000. http://www.who.int/medical_devices/publications/en/Donation_Guidelines.pdf (accessed on August 27, 2013).
15. U.S. Congress. The mercury-containing and rechargeable battery management act 1996, Law 104–142. http://www.epa.gov/osw/laws-regs/state/policy/p1104.pdf (accessed on August 27, 2013).
16. Health and Safety Executive. 2013. The control of substances hazardous to health (COSHH). http://www.hse.gov.uk/coshh/ (accessed on June 2013).

# 13

## Sources of Information for Equipment Management Professionals

### 13.1 Introduction

Equipment management professionals gain their knowledge from education, experience and detailed study of science, engineering, technology and applicable legislation, regulations and guidance. However, laws and regulations are open to interpretation and increasing in number and complexity, and technology is constantly developing and bringing in new technical and clinical knowledge. As new opportunities arise for improving best practice, equipment management professionals need access to a large and increasing body of knowledge. As they encounter new challenges, they will need advice and assistance from others with more advanced knowledge and specialised skills, and they must be prepared to communicate their own novel experience to others. Professionals usefully interact with peers and colleagues in their own and other disciplines, to create a consensus on how best to approach new situations and consolidate and disseminate knowledge to advance clinical practice and professional development.

In this chapter, we introduce the types of agencies and organisations that provide knowledge, advice, guidance and practical assistance to those involved with equipment management, under a number of topics (see a summary in Table 13.1). We explore some key issues, highlight representative bodies and some of the resources they offer, and indicate where a professional might interact with them or contribute to their work. More information can be found on their websites, which are readily located by using an Internet search engine.

**TABLE 13.1**

Types of Organisation Providing Advice and Information on Various
Medical Equipment Management Topics at the National and
International Level

| Advice Provided on | At National Level | At International Level |
|---|---|---|
| Regulation and compliance | Government agencies<br>Approved bodies | Trade bodies and<br>associations |
| Standards and good<br>  practice | Standards bodies<br>Learned societies | ISO, IEC, WHO<br>  Learned societies |
| Medical equipment<br>management service<br>delivery | Consultancy bodies<br>Commercial providers<br>Learned societies<br>Professional bodies | WHO, ECRI, IFMBE<br>Commercial providers |
| Safety and incidents | Non-profit bodies<br>Learned societies<br>Professional bodies | WHO, ECRI<br>Learned societies |
| Education and training | Professional bodies<br>Learned societies | Learned societies<br>Charities |

## 13.2  Government Agencies and Medical Device Regulatory Bodies

### 13.2.1  Regulation

Each national government sets the legal framework under which medical device regulations operate and provides a financial and organisational framework for those bodies which interpret and enforce them. Regulatory bodies oversee compliance with legal requirements, often being responsible for the approval and registration of manufacturers and devices. They can provide the equipment management community with practical guidance on compliance with regulations and on reporting and tracking medical device hazards. Chapter 3 introduces regulatory concepts relevant to medical devices and in this section we highlight some differences in national medical device laws and regulations. For example, many types of equipment are recognised as medical devices internationally, but variation exists between regulators as to what risk classification should be assigned or even if some items qualify as medical devices at all (see Chapter 2). Difference exists, for example, over how to treat aids for disabled people, spare parts and devices incorporating biological tissue. In some jurisdictions, diagnostic devices are covered by separate regulations. Cosmetics may be regulated to various extents through the same body as medical devices or drugs, as is the case for the US Food and Drug Administration, or not at all.

## 13.2.2 Regulatory Bodies

Regulatory bodies seek to ensure compliance with local laws and regulations. They promote the safe design and clinical efficacy of devices, together with their safe use in practice. They provide routes by which manufacturers achieve compliance and usually provide advice and guidance. They clarify ambiguities in the regulations and identify implications for manufacturers and users. Their functions usually include the approval, registration and licensing of devices, including overseeing clinical investigations and placement on the market. They may undertake or oversee inspections of manufacturing plants and quality systems, and accredit bodies that do so, and will oversee or implement hazard reporting and dissemination schemes. Reviews and evaluations of devices on the market may be carried out by a variety of public and commercial organisations that in some countries include bodies acting on behalf of governments and regulators.

Regulatory bodies are often responsible not only for licensing and safe use of devices but also for accrediting other services relevant to medical device regulation, such as test houses and quality system inspection bodies. Related regulatory regimes are likely to cover pharmaceuticals, ionising radiation and health and safety, and in many countries the body that oversees medical devices also oversees drug regulation.

Regulatory frameworks work in similar ways across much of the developed world; the major contrast in the developing world is between those countries that operate a regulatory framework for medical devices and those that have none. The US Food and Drug Administration designates 'tier-1' countries as those where comparable regulatory requirements for medical devices exist, including Australia, Canada, Israel, Japan, New Zealand, Switzerland, South Africa and any member nation in the European Union (EU) or the European Economic Area [1]. Outside this group, medical device regulation is well developed in some countries, partially developed in others, and some have regulations which are not yet legally enforced. In some countries, medical device legislation is neither developed nor enforced [2].

## 13.2.3 Competent Authorities and Notified, Accredited and Certification Bodies

Competent authorities are the agencies in each country responsible for managing the implementation of legislation. For medical devices, the competent authority is usually the national ministry of health or bodies acting on its behalf. In the United Kingdom, the Medicine and Healthcare products Regulatory Agency (MHRA) is the competent authority, established under UK consumer protection law implementing the European Medical Devices Directives. Guidance notes describing how a competent authority designates and works with notified bodies under EU and UK regulation are available on its website [3].

In the EU, a notified body is an organisation appointed by the competent authority to assess whether medical devices and manufacturers meet certain regulatory requirements, with power to issue certificates of conformance. They may assess a manufacturer's quality systems, review a design dossier to ensure a device meets requirements, test devices against technical standards and verify that quality assurance systems deliver consistently compliant products.

Accredited bodies can be public or private sector organisations that accredit products or processes or audit organisations against required standards. For example, in the United Kingdom, the United Kingdom Accreditation Service (UKAS) accredits test and calibration laboratories and accreditation schemes for around 1500 testing laboratories and 500 calibration bodies including medical laboratories, inspection bodies and imaging services. UKAS also accredits certification bodies to perform assessments and award certification against a number of European and international standards regarding product conformity (EN45011), quality management systems (EN45012) and personnel competence (EN45013) [4]. Reciprocal agreements between countries allow international recognition of medical device approvals, compliance with standards and calibration certificates.

## 13.3 Standards and Standards Bodies

### 13.3.1 Overview of Standards

Standards set out a technical description of what something should be or do. Certification to a standard means that a minimum level of performance has been met, which in clinical engineering might mean that an electrical device is constructed safely or that the test regime used to verify this has been carried out correctly. Various bodies devise and publish technical standards of construction, performance and testing which provide a best professional view of what is desirable, achievable and acceptable. Standards are either normative or informative. Normative publications set out a technical description of the characteristics to be fulfilled by the product, system, service or object in question. Informative publications provide background information, such as procedures or guidelines for implementing a standard.

Standards underpin regulation by providing essential detail about a product, process or procedure, based on the expert knowledge and judgement that manufacturers and others follow in their day-to-day practice. One US organisation, the National Electrical Manufacturers' Association (NEMA) (see succeeding text), groups related standards by product or product family, each standard covering one or more of the following elements: nomenclature, composition, construction, dimensions, tolerances, safety, operating characteristics, performance, ratings, testing and the service for which it is designed [5]. Medical device regulations are likely to reference standards on the design, manufacture and testing of medical devices and component parts and also ones covering quality management systems for their fabrication, testing and placing on the market. For example, International Organization for Standardisation (ISO) medical device standards are included as essential requirements in many national medical device regulations.

Regulations and standards are continually being introduced and updated in response to new developments in technology or practice. Professionals must be familiar with the latest technical and process standards relevant to their area, in order to be able to advise correctly on the interpretation of regulation and undertake effectively the design, manufacture, procurement, testing and management of medical devices. Scientists and engineers involved in equipment assessment, equipment modification, development and in-house manufacture also need a detailed understanding of the science and technology behind these standards.

### 13.3.2  Standards Bodies

Many bodies across the world produce standards. These standards may be referred to by regulations, giving them some weight in national law. Standards are voluntary agreements and hence need to be based on a consensus of expert opinion. Participants in their development typically include government body representatives, individual companies and industry bodies involved with manufacturing and testing, academic experts, learned societies and consumer organisations. International standards are agreed by consensus amongst participating nations, each with equal voting rights. The language of international standards is usually English. Once published, a standard remains in use until it is superseded by a later issue or becomes incorporated into or is replaced by another standard.

One of the earliest modern standards bodies was the British Standards Institute founded in 1901. It issues over 27,000 standards, with about 1,700 new ones produced annually. Its committees develop national and international standards, and it is a notified body under the UK medical devices regulations. A different mode of operation is adopted by the American National Standards Institute (ANSI). It does not develop standards itself but provides a framework to enable individual American organisations to develop voluntary standards. It is a private not-for-profit organisation, founded in 1918, that accredits and oversees over 200 public and private organisations and issues 10,000 voluntary and American national standards. One accredited organisation is the NEMA, the largest trade association for the American electrical manufacturing industry, which develops and publishes technical standards, advocates industry-wide policies on legislative and regulatory matters, and collects, analyses and disseminates industry data. Some common quality assurance tests for imaging equipment are based on use of NEMA protocols, including those for gamma cameras and performance testing of ultrasound scanners.

Three global organisations develop international standards in the field of electrical, electronic and related technologies. These are the International Electrotechnical Commission (IEC), ISO and the International Telecommunications Union (ITU). Of these, IEC and ISO produce those standards most relevant to medical technology. The ISO is perhaps the best known internationally and most wide-reaching of the many standards bodies. When appropriate, the IEC cooperates with ISO or ITU. Whereas the IEC and the ITU are concerned with electrical products, ISO standards cover a much wider range of products and activities in the industrial, technical and business sectors. Copies of ISO and IEC standards can be obtained either directly or through national standards bodies, which may also provide advice on their interpretation in a national context. Standards are subject to regular review and revision, for example, ISO standards are reviewed for confirmation, revision or withdrawal 3 years after publication and then every 5 years. International standards developed by ISO may be used directly in some countries or be incorporated indirectly as a national standard in others. Thus, the nearest American equivalent to the CE mark is the UL mark issued in accordance with the essential requirements of NEMA standards. Another example is IEC standard 60601 published by the British Standards Institute as the British standard EN60601 and as UL 2801 in America by NEMA.

Most standards organisations have specific groups working on medical devices–related standards or will rely on the work of other groups for certain subspecialties. One example in clinical engineering is electromagnetic compatibility, where standards rely heavily on work by the Comité International Spécial des Perturbations Radioélectriques (CISPR). This IEC organisation recommends emission limits and threshold levels that ensure the immunity of susceptible devices and specifies appropriate methods of measurement and instrumentation. Standards developed by the CISPR are used extensively throughout the world, and they are an essential contribution to the diverse range of standards for many electrical devices.

Although standards that apply solely to a single country or region are common, there is a continuing tendency to take local standards and adapt them into *harmonised* standards which are more widely applicable. A harmonised international standard is usually based largely on one developed in a particular country which is then amended by standards bodies in other countries to form the basis of their own, locally determined national standard. This can lead to time lags and subtle differences between apparently similar standards. It is important not to assume that compliance with one national standard automatically implies compliance with another similar one unless this is specifically stated by a competent authority or certification body. Choosing the correct version of a standard is of particular concern to manufacturers who supply across international borders, and specialist advice is needed.

### 13.3.3 Key International Standards Relevant to Medical Devices and Their Management

A useful summary of current and expired technical standards specifically applicable to medical devices is available from the EU website [4]. Those most applicable to medical equipment management can be grouped under three headings.

#### 13.3.3.1 Device Construction and Development

Multiple standards address medical device design and construction. Those most useful to start with are the following:

- *IEC 60601* – the first international standard directly relevant to medical devices with over 160 individual component standards, clustered under three headings: general requirements, particular requirements for specific device types and requirements for performance testing specific device types [6]
- *ISO 14971* – addresses the management of risk during the design, construction and delivery of medical equipment [7]
- *ISO 14155* – aims to help medical device companies conduct clinical trials [8]

#### 13.3.3.2 Quality Systems

These are fundamental in achieving compliance with medical device legislation and regulatory requirements and for manufacturing medical devices in the EU. They also support effective in-house clinical engineering services, and the UK MHRA recommends that all medical equipment maintenance services are certificated to such standards [9]. Quality systems common in clinical engineering include the following:

- *ISO 9000 series and ISO 9001* – set out a generic quality system standard relevant to all equipment management processes including design, production, maintenance, repair, procurement and project management [10,11].
- *ISO 13485* – similar in scope but contains requirements particularly applicable to medical devices. It defines process standards for medical device manufacture, so is relevant to in-house development and modification of devices [12].

- *ISO 17025* – a quality management standard aimed at demonstrating that test and calibration laboratories produce technically valid results. Clinical engineering departments may seek accreditation to this standard when providing a test or calibration service for equipment diagnostic and treatment [13].

#### 13.3.3.3 Medical Equipment Management

Standards supporting this include the following:

- *ISO 10012* – looks at how to manage measurement equipment [14]
- *IEC 60812 [15] and IEC 61025 [16]* – set out techniques for analysing system reliability and faults
- *IEC 62353 [17]* – defines requirements for the frequency and nature of routine electrical testing of medical equipment

### 13.3.4 Other Functions of Standards Bodies

Standards bodies do not only develop standards. They may also perform or oversee certification and testing and audit services seeking conformity to their standards. They provide services and resources in the areas of standardisation, best practice guides, testing and certification and training and audit. They may also provide training and information on standards and international trade. Some national standards bodies take an active role in promoting and developing their standards for international use. For example, the ANSI promotes US standards internationally through the ISO and the IEC, of which it administers many key international technical committees and subgroups, whose members comprise volunteers from industry and government. A practising clinical engineer will find all these bodies to be useful sources of advice and guidance, especially where their professional body works closely with their national standards organisation.

## 13.4 Learned Societies and Professional Bodies

### 13.4.1 Introduction

Membership of one or more peer-based groups is an essential component of what it means to be a professional in any field. It helps individuals achieve recognition of professional status, keep up with developments and satisfy requirements for continued practice. There are two broad types:

1. Learned societies, which are uni- or multidisciplinary bodies promoting a particular subject or area of clinical practice or technology. They provide detailed practical advice on implementing

legislation, standards and best practice and are an important source of information and ideas that advance professional practice and techniques.

2. Professional bodies, which are directed mainly at supporting and developing a group of professionals in a particular discipline. They often incorporate a learned society function and may also provide or oversee professional training or registration and represent members in developing local and national policy.

Professional bodies often support the provision of training and may accredit its delivery or assessment processes where these are relevant to entry into their profession. They may also award additional qualifications or other recognition of successful individual progression to a senior level, such as chartered or incorporated membership, and fellowship for those making an outstanding contribution to their field. Individuals may belong to more than one learned society, depending on their professional interests, but tend to stay with one professional body. Grade of membership will not usually restrict access to sources of learning and advice. Membership of a society, even at associate level or with a group outside a person's immediate professional area of practice, brings resource benefits and can be a valuable source of learning and development.

Clinical and healthcare practitioners have access to a wide range of national and international societies and professional bodies that cater for their specific interests and provide targeted publications, educational events and meetings accessible to their members and others.

Professional bodies and learned societies generally have few paid employees, and their primary functions are to administer the society rather than act as scientific or technical experts. There is great scope for an individual to extend their professional and individual development through participation in the work of these societies, whether organising training and scientific meetings or serving on committees, panels and special interest groups. Such activities will be unpaid, although expenses are usually reimbursed. They can take up a significant proportion of time for senior professionals but are of general value to the disciplines being supported and provide continuing personal development for the participants. Patients and the public are increasingly seen as a valuable resource to medically orientated groups in helping to direct training and set priorities in resource use and allocation of research funding.

Professional bodies can provide some of the training required for postqualification career markers, together with continuous professional development through courses, meetings and publications related to clinical practice. This development is recognised by professional qualifications and awards, such as incorporated or chartered engineer. Societies usually separate professional from trade union functions that directly represent the financial or employment interests of their members, because professional bodies and learned societies can gain advantages from holding charitable status such as tax exemption and the ability to raise funds by public donation. In the United Kingdom, charitable status can be achieved by organisations whose aims include the advancement of education, science or health and increasing public consciousness of their subject; lobbying actions in both public and political circles are acceptable if directed towards these ends. Charitable body restrictions vary between countries.

### 13.4.2 Societies Relevant to Equipment Management and Healthcare Technology

A large proportion of clinical and service professionals in healthcare use medical technology and many manage it. Their professional bodies and learned societies usually provide some education and training relevant to equipment use and management. Professions where medical device issues are prominent include anaesthetists, radiologists and intensivists, to give just three examples, and the scientific, engineering and support staff who support them. Societies keep members up to date with new technological and device developments and their associated clinical, safety and legal implications. In addition, they may provide assistance with device procurement by providing, for example, standard specification forms for use when tendering for imaging equipment, such as those provided by the British Medical Ultrasound Society and the British Nuclear Medicine Society for different types of imaging equipment (see Chapter 6). Societies can also provide advice on the planning and installation of equipment in specialised areas, and the European Intensive Care Society, for example, publishes minimum requirements for intensive care units.

Societies set up to support equipment-intensive areas of medicine include those devoted to clinical engineering and medical physics (separately or in combination) and intensive care medicine and those including anaesthetists, intensivists, surgeons, radiologists and radiographers. All those concerned with imaging have a close interest in the nature and performance of their equipment and the way in which it is used. To illustrate that equipment management is not just the province of clinical engineers, consider medical physicists, who play a major role in the specification, procurement, commissioning and testing of imaging and therapy systems using ionising and non-ionising radiation. They are closely involved in risk and safety management and may deal with the interconnection of systems for image and data acquisition, storage and processing, including gating using physiological signals. Clinical engineers and related societies, often involve and educate those from other professional groups in equipment issues as they work alongside them to deliver services to patients.

The major UK-based organisations that represent core professions related to clinical engineering and medical physics in healthcare include the Institute of Physics and Engineering in Medicine (IPEM), the Institute of Engineering and Technology (IET) and the Institution of Mechanical Engineers (IMechE). Other relevant UK professional associations include the Institute of Healthcare Engineering and Estate Management (IHEEM) and the Chartered Institute of Public Finance and Accountancy (CIPFA) for those who approach equipment management from a financial, procurement or service management direction. The US-based Institute of Electrical and Electronics Engineers (IEEE) is by far the largest international technical organisation, with over 400,000 members in more than 160 countries. The IEEE has active subgroups devoted to medical science and technology which publish journals such as the *IEEE transactions* series and regularly organise conferences and meetings.

### 13.4.3 Training and Certification in Clinical Engineering

Requirements for the accreditation and registration of clinical engineers and technologists vary between countries and sometimes between different healthcare organisations in the same country. Part of the work of a clinical engineer or technologist may also overlap with a role adopted by another professional group with a different set of requirements for training and qualification. Registered health care providers must demonstrate the competence of their employees, and transferable qualifications and marks of achievement in medical equipment management are desirable. Education to degree or postgraduate level coupled with vocational training is likely to comprise the main international route for developing clinical engineers and technologists. National professional bodies may provide or oversee vocational or specific training in clinical engineering.

Globally, a number of societies are directly concerned with the accreditation of engineers. The American College of Clinical Engineering (ACCE) is a professional society representing the interests of clinical engineers which aims to establish standards of competence and promote excellence in clinical engineering. Membership requires 3 years of professional practice of engineering in a clinical environment, coupled with a suitable academic qualification, which is usually a degree in engineering. In the United States, certification in clinical engineering is under the administration of the Healthcare Technology Certification Commission through its US Board of Certification in Clinical Engineering. This programme is supported by the ACCE Healthcare Technology Foundation. Its certification process includes testing for a minimum level of knowledge through written and oral examinations and a minimum level of continuing education for renewal. ACCE certification is not restricted to the United States and the examinations can be taken in most major cities around the world. Although certification is voluntary, it is recognised by many employers as a measure of appropriate professional accomplishment.

In Canada, certification is available for biomedical engineering technicians, individuals who directly support the service and repair of hospital equipment, from an accreditation board (CTAB). Biomedical engineering technician education and training is of a more directly technical nature than that of clinical engineers and is supplemented with specific training to support medical equipment. This qualification, Certified Biomedical Engineering Technician/Technologist (CBET), is recognised across North America alongside related certifications such as Certified Laboratory Equipment Specialist (CLES) and Certified Radiology Equipment Specialist (CRES). Certification requires a combination of experience, referee support and passing a written examination. For biomedical technicians, the syllabus requires knowledge of anatomy and physiology, electronics, medical instrumentation, troubleshooting and standards and practice. Professional bodies can provide materials and other support to help individuals achieve accreditation.

### 13.4.4 Knowledge Resources

Professional bodies and learned societies publish reports and guidance for their members and others, usually at a reasonable cost. The advent of the Internet has made it much easier to find out what is available, including specialist courses and conferences. In the UK fields of clinical engineering and medical physics, the IPEM arranges meetings and courses and produces reports on many aspects of medical device operation, quality assurance and calibration together with a number of journals that highlight developments in these disciplines. Many articles are concerned with practical measurement and quality assurance, and specific engineering reports include the safe design of electromedical equipment [18] and notes on medical devices regulation. Other societies mentioned in this chapter produce similar resources and increasingly offer online courses and self-directed learning. We encourage readers to contact colleagues and search actively for information through professional bodies and other groups.

### 13.4.5 Regional and International Societies

Regional and international groupings of national societies are usually developed to address particular supranational groupings or to support countries facing similar problems. The European Federation of Organisations in Medical Physics (EFOMP), a Regional Organisation of IOMP, was founded in 1980 and now comprises 35 national organisations that represent more than 5000 physicists and engineers working in medical physics. Amongst its aims are to develop guidelines for education, training and accreditation programmes, to make recommendations on the responsibilities and roles of medical physicists, and to work for Europe-wide recognition of medical physics as a regulated healthcare profession in all member states. Its clinical engineering equivalent is the European Alliance for Medical and Biological Engineering and Science (EAMBES). Some regional societies enrol individual members and operate as learned societies, but most have national societies as their individual members.

International societies provide a global focus for their specialty and provide a forum in which to reach international consensus on key issues when dealing with other international organisations such as the World Health Organization (WHO) and the United Nations (UN). Like national societies, international bodies can represent a profession, a subject area, or both. The International Federation for Medical and Biological Engineering (IFMBE) is an association of constituent societies and national organisations, established in 1959 to encourage and promote international collaboration in research, practice, the management of technology and the application of science and engineering to medicine and biology. The Federation has ambitious and wide-reaching aims, seeking to influence the practice of medical engineering at a global level and to raise both professional and public awareness of the discipline. It recommends policies and provides guidelines in professional, educational and ethical areas. The IFMBE has a Clinical Engineering Division focusing specifically on healthcare technology management with its own programme of activities. The IFMBE publishes proceedings book series covering the various topics of medical, biological and clinical engineering, together with a series of books intended for both students and researchers that covers many aspects of clinical engineering and a major journal, *Medical and Biological Engineering and Computing*. IFMBE is recognised by both the WHO and the UN, through its membership of the International Union for Physical and Engineering Sciences in Medicine (IUPESM), as covering the full range of biomedical and clinical engineering, healthcare and healthcare technology management. Through its 50 national and international member societies, it represents more than 120,000 professionals involved in predominantly engineering issues to improve healthcare delivery. IFMBE cooperates with the WHO in the fields of medical, biological, clinical and hospital engineering, management of biomedical technology and technology development, assessment and transfer, resource building and patient safety. Every 3 years, the IFMBE holds a world congress on medical physics and biomedical engineering in cooperation with the International Organization for Medical Physics (IOMP). The IOMP was founded in 1980 and provides the same type of international forum for its 82 member societies in the field of medical physics.

Similar structures apply to national and international societies devoted to the study of applications of a particular technology, such as medical ultrasound. The American Institute of Ultrasound in Medicine has 15 networking communities, each of which covers a particular aspect of medical ultrasound, and belongs to a global umbrella organisation, the World Federation for Ultrasound in Medicine and Biology (WFUMB). Similarly, there are over 20 national societies that represent the discipline of nuclear medicine that are represented by the World Federation of Nuclear Medicine and Biology (WFNMB). Single international societies also exist, such as the International Society for Magnetic Resonance in Medicine (ISMRM), where over 8000 professionals from over 50 countries tap directly into a global network.

### 13.4.6  Chartered Status

Professional institutions can confer chartered status on individuals who meet suitable criteria. A chartered professional is a person who has gained a recognised level of competence in a particular field of work. Historically, this chartered status was awarded by UK institutions that had been incorporated under royal charter by the British monarch, but many organisations now issue chartered designations without royal or parliamentary approval. As a practice in the engineering professions, it is common in Britain and has been adapted by similar organisations around the world. One benefit of chartered status is that it is usually recognised outside the specialty represented by membership of a particular society. Chartered status in engineering (CEng registration) is usually obtained through a professional engineering institution licensed for the purpose by the Engineering Council in the United Kingdom, such as the IPEM, the IET and the IMechE. Other chartered professionals working with healthcare technology in the UK include chartered physicists (CPhys) and chartered scientists (CSci).

The competence requirement for a chartered engineer working with healthcare technology is an appropriate scientific and clinical background, plus relevant specialist education, knowledge and experience, together with problem solving, managerial, budgeting and organisational skills. They should be able to find solutions to day-to-day engineering problems and develop and apply novel technologies, engineering services and management methods. Chartered engineers must possess leadership and effective interpersonal skills and act with honesty and integrity. The ability to show effective leadership and produce successful engineering outcomes is key to chartered status. Possession of this status is an important indicator of an individual's suitability for a demanding senior clinical engineering role. Incorporated engineer (IEng) and EngTech are alternative types of registration which confer similar professional benefits to CEng to different groups. A clinical engineering professional should aim for state or voluntary registration, chartered (or incorporated) status and membership of a professional body.

## 13.5  Sources of External Assistance: Commercial, Non-Profit and Peers

### 13.5.1  Commercial Equipment Advisors and Equipment Manufacturers

Consultancy services and advice are available from a variety of commercial, government, non-profit and peer-review bodies. This can involve benchmarking, audit, review of services, provision of infrastructure such as software, testing of devices and investigation of adverse incidents. These bodies may employ experienced clinical engineers, working on a part-time or consultancy basis, who make use of knowledge and expertise gained through service in the industry. Similarly, consulting companies are available, which can advise on or undertake specific equipment management projects that range from complete major new build PFI and public–private partnership projects to production of equipment inventories, assessing service delivery models and establishing equipment replacement programmes. Such companies can be identified and contacted through the many printed and online trade directories published throughout the world.

It is natural to look to equipment manufacturers for support in training, maintenance and general information in connection with their own products. However, some provide a number of further services in the areas of procurement, multivendor maintenance, training in specialist technical skills and third-party managed services. These include maintenance contract bundles which are offered both by manufacturers, who are now often found maintaining and supporting equipment other than their own, and by specialised independent providers. Third-party maintenance is a growing area (see Chapter 8).

Some companies organise conferences at which experts and users can present their experiences of using products in routine or novel clinical settings or provide technical overviews of the principles behind new and existing devices and techniques. Online forums provide sources of practical information for technicians who repair, maintain and support equipment, and discussion with professional peers will indicate which are the most useful and reliable.

An important area of involvement for clinical engineering professionals is innovation, where collaborations between academics, hospitals and manufacturers in the development and assessment of novel device or techniques allow these to move from the laboratory to clinical use. The specialist technical expertise and test resources of companies can also be helpful in incident investigations, particularly where these are carried out in collaboration with national regulatory bodies.

### 13.5.2 Supranational and Not-for-Profit Agencies

Supranational organisations include international standards bodies, groups of professional bodies and societies and large charities. These act to promote good practice and develop guidance, scientific and technical knowledge and professional issues at a global level. Several supranational organisations and charities operate internationally and provide resources of great practical assistance to clinical engineers around the world.

The WHO is part of the UN. It promotes the development of regulatory systems for medical devices worldwide and is responsible for, 'Providing leadership on global health matters, shaping the health research agenda, setting norms and standards, articulating evidence-based policy options, providing technical support to countries and monitoring and assessing health trends' [19]. The WHO is particularly valuable to those concerned with providing healthcare technology management in developing countries and to the growing number of charities and professional volunteers working in these regions. It is of assistance to manufacturers, importers, exporters and innovators and provides a wealth of information on local regulations and regulatory bodies. It has produced a number of comprehensive and informative guidance documents that advise on aspects of healthcare technology management, including medical device regulation [20,21].

The WHO has regional and country offices worldwide. Its diagnostic imaging and medical devices unit coordinates the Global Initiative on Health Technologies. It works with nongovernmental organisations (NGOs) such as the IFMBE, IOMP, Health Technology Assessment International and the International Network of Agencies for Health Technology Assessment. Resources produced by the WHO include reference documents and training modules in the areas of needs assessment, procurement processes, donation, inventory management, maintenance and computerised equipment management systems. The WHO is particularly concerned to keep abreast of what devices are currently available and the ways in which they are used and to identify areas where developments in usage and research are required [22].

### 13.5.2.1 ECRI Institute

The ECRI institute is an independent not-for-profit research agency set up in 1968 to provide information, training, education and consultation related to medical devices. It provides medical equipment management support services worldwide. Using its test laboratory and extensive databases, it can provide advice on everything from specifications to life-cycle cost analysis, analyse and interpret terms and conditions of contracts, perform tendering and bid analysis and advise on purchasing and financing. It can also contribute to process analysis and improvement. It is able to provide advice on emerging technologies, product specifications, hazards and recalls, user experience, clinical guidelines, standards and regulations and data on medical device performance. It also produces, licences and supports its own equipment management database.

### 13.5.2.2 Charitable Bodies

The personal satisfaction of contributing to the work of a charity brings its own benefits. However, there are also professional benefits for the engineer or clinician who becomes involved, providing an opportunity to work with unfamiliar equipment or healthcare systems, often under challenging conditions with limited resources. This develops interpersonal and management skills, broadens technical knowledge and is a valuable part of continuing professional development. Examples of charitable organisations involving clinical engineers include Medecins Sans Frontieres (MSF), which provides medical aid in emergency situations throughout the world; Engineers Without Borders (EWB), a UK NGO that aims to promote the use of engineering for human development; and Mercy Ships, which sends ships equipped with operating theatres to countries in the developing world to perform free surgery. Other groups include the Chain of Hope and Engineering World Health (EWH) and there are many more.

## 13.6 Summary

The amount of information needed by a professional to practise effectively and safely is increasing in scope, volume and complexity. To keep up with all this information is challenging. For example, developments in regulations and standards mean individuals need assistance and support on their interpretation and implementation. Help is often available from the regulatory and standards bodies themselves, and from consultants and advisors in both the commercial and not-for-profit sectors. Professional bodies and learned societies play a part in training and professional development, sometimes directly contributing to the processes by which professional qualifications are obtained, and taken together they provide a vast learning resource of publications and meetings. They also provide practical help, for example, producing guidelines on how equipment-intensive clinical units are set up or in producing templates for purchasing specifications. There is scope for the professional to contribute to many of these organisations, something which enhances an individual's knowledge and widens skills. Opportunities include serving on professional and standards body committees and contributing to charitable efforts. Some practical steps for the clinical engineering professional wanting to become more aware of and involved in the wider context to their profession include the following:

- Find out who your national regulator is and how they work.
- Identify the competent authority, read their guidance and advice and develop contacts for clarification when needed.
- Understand the content and background to relevant standards, and keep up to date as these are revised in the light of new technical knowledge and practice.
- Contribute to the development and updating of standards and good practice.
- Use standards bodies, professional bodies and learned societies for advice and guidance.

- Search for and learn from available reference and teaching material.
- Link up with like-minded professionals and develop yourself and your profession.
- Consider how best to involve patients and the public in your own practice.

## References

1. US Food and Drug Administration. Import and export of investigational devices. http://www.fda.gov/MedicalDevices/DeviceRegulationandGuidance/HowtoMarketYourDevice/InvestigationalDeviceExemptionIDE/ucm051383.htm (accessed on September 05, 2013).
2. World Health Organization. Medical devices – Medical devices regulations. http://www.who.int/medical_devices/safety/en/ (accessed on September 05, 2013).
3. MHRA. How we regulate. http://www.mhra.gov.uk/Howweregulate/index.htm (accessed on September 05, 2013).
4. European Commission. European standards – Medical devices. http://ec.europa.eu/enterprise/policies/european-standards/harmonised-standards/medical-devices (accessed on September 05, 2013).
5. NEMA. About NEMA standards. http://www.nema.org/Standards/About-Standards (accessed on September 05, 2013).
6. IEC 60601-1 Ed 3.1. *Medical Electrical Equipment – Part 1: General Requirements for Basic Safety and Essential Performance*. International Electrotechnical Commission, Geneva, Switzerland, 2012.
7. ISO 14971: 2012. *Medical Devices – Application of Risk Management to Medical Devices*. ISO, Geneva, Switzerland, 2012.
8. ISO 14155:2011. *Clinical Investigation of Medical Devices for Human Subjects*. ISO, Geneva, Switzerland, 2011.
9. MHRA. DB2006(05). *Managing Medical Devices*. Reviewed in 2013. Medicines and Healthcare Products Regulatory Agency, London, U.K., 2006.
10. ISO. ISO 9000. Quality management. http://www.iso.org/iso/home/standards/management-standards/iso_9000.htm (accessed on September 05, 2013).
11. ISO 9001:2008. *Quality Management Systems – Requirements*. ISO, Geneva, Switzerland, 2008.
12. ISO 13485:2012. *Medical Devices – Quality Management Systems – Requirements for Regulatory Purposes*. ISO, Geneva, Switzerland, 2012.
13. ISO 17025:2005. *General Requirements for the Competence of Testing and Calibration Laboratories*. ISO, Geneva, Switzerland, 2005.
14. ISO 10012:2003. *Measurement Management Systems – Requirements for Measurement Processes and Measuring Equipment*. ISO, Geneva, Switzerland, 2003.
15. IEC 60812. *Analysis Techniques for System Reliability – Procedures for Failure Mode and Effects Analysis (FMEA)*, 2nd edn. International Electrotechnical Commission, Geneva, Switzerland, 2006.

16. IEC 61025. *Fault Tree Analysis (FTA)*, 2nd edn. International Electrotechnical Commission, Geneva, Switzerland, 2006.

17. IEC 62353 ed1.0. *Medical Electrical Equipment – Recurrent Test and Test After Repair of Medical Electrical Equipment.* International Electrotechnical Commission, Geneva, Switzerland, 2007.

18. Institute of Physics and Engineering in Medicine. *Report 90: Safe Design, Construction and Modification of Electromedical Equipment.* IPEM, York, England, 2004.

19. WHO. About WHO. http://www.who.int/about/en/ (accessed on September 05, 2013).

20. WHO. *Medical Device Regulations: Global Overview and Guiding Principles.* The World Health Organisation, Geneva, Switzerland, 2003.

21. WHO. *A Model Regulatory Program for Medical Devices: An International Guide.* Pan American Health Organisation, Washington, DC, 2001.

22. WHO. *Medical Devices: Managing the Mismatch: An Outcome of the Priority Medical Devices Project.* The World Health Organization, Geneva, Switzerland, 2010.

# 14

## Improving Performance: Quality, Indicators, Benchmarking and Audit

### 14.1 Introduction

Monitoring is integral to the success of any productive activity, whether detailed inspection of each step in a process or general user feedback on the outcome. Since monitoring has a cost, it makes sense to measure only what is essential to control and improve. There are also wider reasons to measure performance in an organisational context including providing evidence that financial targets and governance rules are being met.

Healthcare systems involve multiple interests. Their performance is judged not just on financial grounds or clinical outcomes but also by broader measures including quality, choice and user empowerment. To monitor the performance of a healthcare organisation requires asking questions such as: 'How well are our services delivered?', 'Are our patients/clients satisfied?', 'Are outcomes what they should be?' and 'Are we achieving our financial and strategic aims and objectives?' Healthcare has a high political profile, and in publicly funded or coordinated systems, politicians are under pressure to achieve value for money and demonstrate they are introducing improvements. Every healthcare organisation must aim to show it meets financial, governance and performance requirements, is popular with its patients and can stand up to the scrutiny of politicians and the public. Continued support depends on maintaining a good reputation, and demonstrating good performance influences this through pressure groups, the media and status in comparative league tables.

In this chapter we outline the drivers for performance measurement and improvement. We consider what makes a good performance indicator in clinical engineering and how these may be devised to produce productive changes in organisational behaviour, whilst avoiding perverse incentives. We look at how benchmarking can improve services and finally consider how to use audit to improve organisational performance.

## 14.2 Why Monitor Performance?

There are four general reasons to monitor ongoing performance:

- *Process control*: Service or product delivery depends on the reliable operation and interaction of multiple processes. Objectively monitoring a process and its output gives managers the information to control it – the old saying, 'If you cannot measure an activity you cannot manage it', has been widely accepted for many years [1]. Suitable performance indicators can monitor effectiveness (the ability to get things done) and efficiency (getting them done with a minimum of resource), and pick up and help to control problems.
- *Output monitoring*: Outputs need checking against their specification, as an organisation which does not deliver the right product or service on time is unlikely to remain in business for long.
- *Service quality*: Performance indicators help to maintain and improve service quality. These include qualitative measures such as customer surveys and broader feedback from interested parties, including clients, customers, funders and organisational management.
- *Reassuring stakeholders*: Suitable indicators demonstrate that an organisation has control of its operations. They show senior management, patients, clients, customers and external bodies whether the service is monitoring and improving its processes, outputs and outcomes.

## 14.3 Internal and External Monitoring

### 14.3.1 Internal Monitoring

Good performance means an organisation is efficient and effective at what it does and that it delivers consistently the right result, at the right time and in the right way. This is the essence of quality. It requires continuous delivery of, 'The totality of features and characteristics of a product or service that bear on its ability to satisfy stated or implied needs' [2]. Internal and external monitoring measures help show whether this is the case, such as testing production output and surveying service customers. Checks also help to identify and correct problems and indicate how to improve working methods and outcomes [3].

Alongside technical performance, an organisation also monitors its performance to provide longer-term assurance to its customers that it will deliver what was agreed. Quality assurance is: 'All those planned and systematic actions necessary to provide adequate confidence that a product or service will satisfy given requirements for quality' [2]. For example, regular internal audits tell an organisation how closely it is following its own policies and procedures, and audit by an external body provides reassurance to patients, managers and other stakeholders that the organisation is working as intended. Quality management systems provide an integrated and proactive approach to monitoring and improvement, through a focus on customers supported by management review, performance indicators and audit. These systems simultaneously support internal improvement and meet external demands for validation. Chapter 3 provides a short introduction to quality management systems.

### 14.3.2 External Monitoring

Multiple external bodies monitor, inspect or review healthcare organisations. Their activities include informal observation, formal inspection, audit and specific demands for information. They may review adverse incidents or indirect information such as patient satisfaction measures before deciding whether or not to inspect an organisation in depth. External monitoring generally benefits an organisation in the long term, as it sharpens self-monitoring and provides valuable external feedback and ideas to drive forward improvements. External bodies that carry out or influence performance monitoring include the following:

- *Government and regulatory bodies* interested in compliance with the law and regulations: For example, the UK Care Quality Commission inspects healthcare facilities against Outcome 11 which requires medical equipment to be 'Properly maintained, suitable for its purpose and used correctly' [4]. This is written in general terms but is supported by detailed guidance [5].

- *Government-sponsored bodies* concerned with clinical outcomes, patient experience, value for money, risk management or progress with healthcare initiatives.

- *Insurers* looking for evidence that risk reduction measures have been implemented to reduce the scale or impact of claims for compensation or damage. For example, the UK NHS Litigation Authority reviews aspects of safety and risk in equipment management [6].

- *Groups and individuals* seeking to drive improvements and support changes in practice: Including charities, patient and political pressure groups and non-government agencies.

- *Self-assessment and peer review* schemes vary from informal professional review to national schemes supporting regulatory action. For example, the Australian Council on Healthcare Standards accredits against a set of standards, self-assessment and regular peer review, and the Joint Commission in the United States undertakes regular inspections, to accredit hospitals and laboratories which are then referred to by most states when licensing healthcare facilities.

### 14.3.3 Monitoring Methods

Monitoring methods include a mixture of self-assessment, internal audit and external inspection. For example, individuals may assess their own performance, have this reviewed by their manager, and report their continuing professional development to a registration body. Departments and services assess themselves by analysing their performance statistics and performing internal audits, whilst external bodies may perform inspections to check that accreditation standards are maintained. Likewise an organisation providing healthcare will monitor itself through its management board and specialised committees but will be assessed by external bodies, acting on behalf of the state, to ensure that it is fit to provide healthcare. Inspections or audits by external bodies may take place regularly, including quality system audits, spot checks and periodic reviews or be triggered by an incident report. Some regulatory bodies do not undertake site visits but use published figures or information specifically requested from an organisation.

## 14.4 Constructing Performance Indicators

Performance indicators are based on measurements carried out under defined conditions and consistent criteria. Indicators provide a basis for monitoring changes in performance and assessing progress in attaining targets. They also support comparisons between services. Quantitative indicators usually have a reference value and range which they can be monitored against over time or between different operations. Qualitative indicators can include, for example, a flag triggered by a specific event.

### 14.4.1 Types of Performance Indicators

Three general types of indicator are used internationally, based on three dimensions of quality originally described by Donabedian [7,8]: structure, process and outcome. *Structural measures* indicate how well an organisation provides the physical and organisational infrastructure and resources to ensure customer requirements are met and external rules are followed. According to Donabedian, *'It refers to the attributes of the settings in which providers deliver health care, including material resources, human resources and organizational structure'* [7]. Appropriate flags need to be identified to show up particular risk thresholds, in areas such as clinical outcomes, medical devices or health and safety. These include how well financial, operational and compliance controls are working. Sources of monitoring information include internal risk assessments, self-assessment against standards and controls, incident reports, complaints and claims information, internal and external audit results (including clinical audit) and inspection visits. Once a risk is identified, the organisation needs to decide whether and how often to monitor it and who should be responsible. Quantitative indicators provide more objective measures which flag variances in process or outcomes for early attention.

*Process measures* include timeliness, responsiveness and proficiency. Timeliness shows whether a service was delivered to the customer within an agreed timescale. Responsiveness is the ability to respond to altered customer requirements, such as a change in specification. Proficiency measures show whether a process was performed correctly. All are affected by how effectively staff work and the time and tools that they have at their disposal.

*Outcome* looks at how well things turn out, for instance, how the infection rate changes on an intensive care unit after hand gel use increases due to more forceful hygiene campaigns and training. Outcome is more difficult to assure than structure or process, because poor outcomes usually result from a combination of factors. Problems often result from unanticipated interactions between isolated faults that combine with other organisational failings to create a chain of events that ends in an accident. Systems may fail due to a lack of integration, where individual subsystems are performing well but there is a fault at the boundaries between them. Constructing indicators to pick up all potential faults in complex systems is difficult and costly. An alternative approach is to have sufficient measures to pick up the most likely problems, whilst adding extra ones to enable the underlying causes of less common faults to be diagnosed so that action can be taken to prevent a recurrence.

### 14.4.2 Quantitative Input and Output Measures and Ratios

Detailed performance indicators can be derived by comparing variables at different stages in service or product delivery, with the ratios showing how effectively resources are used. Common measures fall into the categories of economy, efficiency, cost-efficiency and effectiveness, and Figure 14.1 shows how these interrelate. These can be summarised as follows:

- *Economy*: Measures input resources, such as money, time and materials. It is the easiest measure to make but the least helpful in determining how well a service is working.

- *Efficiency*: The ratio of output to input resource. It is more meaningful than efficiency alone. Measures include the number of jobs done in a day. *Cost-efficiency* compares output with input costs, for example, dividing the number of jobs done by total wages and overheads to give an average cost per job. It is a useful measure of overall service performance.

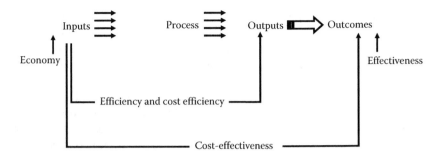

**FIGURE 14.1**
Interrelationship of different types of process measure.

- *Effectiveness*: Considers how well one or more outcomes have been achieved. This can be difficult where there are mixed results – for example, is rapid return of a partially repaired piece of equipment to service more important than a delay to fit new spare parts? *Cost-effectiveness* is the ratio of input to outcome and is a good indicator of service value.

A service manager is constantly balancing effectiveness and efficiency. For example, it is more expensive to pay a courier to deliver a spare part bought at list price in order to quickly repair a breakdown, but if it results in a crucial clinical service being back on stream in an emergency, it is certainly effective.

Finally, *rate-based* indicators are suitable for events which occur at relatively high frequency and lend themselves to statistical analysis. An example is the number of equipment failures per year per thousand units in operation. Statistical process control extends this idea to monitor variations in volume activity [9,10], for example, to flag up when variations in regular image quality tests are too high.

## 14.5 Performance Indicators in Equipment Management

### 14.5.1 Devising Practical Indicators

Not everything in a service needs to be monitored. Deciding what to check depends on the purpose for which the information will be used. This can include checking process performance and integrity, monitoring costs and efficiency, investigating problems, reporting compliance to required standards and identifying ways to improve detailed processes or whole systems. Before devising local performance indicators, it can be productive to see what measures are used by other organisations. Standard indicators are likely to be useful when comparing performance between organisations, whereas local indicators should answer specific needs such as service-level agreement targets and management monitoring and are likely to develop over time [11]. Effective performance indicators are as follows:

- *Important or significant*: Address activities high in cost, priority, risk or volume
- *Valid and attributable*: Based on evidence and clear responsibility for improvement
- *Useful*: Relevant to practice, identifying areas where improvement is possible
- *Definable and measurable*: Able to quantify changes in performance
- *Available*: Easy, quick and cheap to collect
- *Reliable*: Accurate, reproducible, not skewed by random variation or measurement error
- *Meaningful*: Relate to activity and to service-level agreement measures
- *Communicable*: Easily understood by those generating and interpreting them
- *Comparable*: Consistent definitions when benchmarking with other organisations

More than one indicator might be needed for some activities, to avoid unintended consequences. For example, even if a target to repair 75% of broken equipment within 3 days is met, some important items may not be assessed for some months, if at all, with probable customer dissatisfaction. A second indicator might specify that all items be assessed and have their status reported to the customer within 2 weeks. Measurement frequency can be flexed with experience, or to monitor the effects of a change, for example, to see if a new maintenance provider is servicing equipment correctly under a contract. Popular indicators include those concerned with resources use, employee productivity and timeliness [12] such as the following:

- *Timeliness of maintenance*: Proportion of planned maintenance carried out on time
- *Management to staff ratio*: An overhead cost measure which needs careful analysis, as senior engineers often undertake both management and technical functions
- *Employee turnover*: A measure of service stability and a proxy for staff satisfaction
- *Units per person-hour*: An efficiency measure, where a valid comparison might be between two people doing a similar profile of work
- *Earned man-hours*: Efficiency measure of time logged on active work
- *Stock ratio*: Value or quantity of spares held versus those used in a month, to monitor how well ordering matches use

**TABLE 14.1**

Examples of Structural, Process and Outcome Performance Indicators for Various Elements of the Equipment Management Life Cycle

| Area | Structural Measures | Process Measures | Outcome Measures |
|---|---|---|---|
| Assets | • Accurate inventory | • Ability to locate assets | • Replacement plan |
| Maintenance | • System to clearly identify maintenance needs | • In-house maintenance costs as % of inventory value<br>• % of staff time spent on maintenance | • Equipment downtime<br>• % of planned maintenance completed |
| Training | • Policy and procedure agreed to by users | • % of required training completed | • % of repair calls due to user errors |
| Incidents | • Investigation procedures in place | • Levels of reporting | • Frequency of similar incidents |

Clinical engineering services should seek to monitor performance across the whole equipment management process, in collaboration with other departments responsible for risk and finance, to improve internal performance and raise issues of concern to the organisation. In addition to equipment repair, this might, for example, include monitoring capital spend or adverse incidents. Table 14.1 illustrates possible structural, process and outcome indicators in different areas and Table 14.2 highlights some indicators used in group benchmarking. Structural scores estimate the degree to which the criterion is met.

### 14.5.2 Pitfalls of Indicator Design

Performance indicators need to be chosen with care; otherwise individuals tend to respond by trying to optimise only what is being measured and lose focus on quality. Two important questions need to be asked of any performance indicator: 'Does it have the potential to induce desirable change?' and 'Can it lead to perverse incentives?' Perverse incentives are ones that lead to improved performance within the definitions of the indicator but which are detrimental to activities elsewhere that are not being measured. They can also give rise to gaming, where actions are taken merely to comply with the strict wording of a target – such as starting an equipment repair within the declared timescale but immediately putting it to one side, or including only direct labour costs and omitting management overheads.

An example of a robust measure to assess how well a routine maintenance operation is keeping up to date would be 'proportion of scheduled maintenance jobs completed within the specified month'. A similar sounding measure, 'number of equipment repairs completed in a month', may however

**TABLE 14.2**

Example of Benchmarking Clinical Engineering Services between Different Hospitals

| Ref. | Type of performance indicator | Hospital (Anonymised) | | | | | | | | |
|---|---|---|---|---|---|---|---|---|---|---|
| | | A | B | C | D | E | F | G | H | I |
| | **Systems** | | | | | | | | | |
| S1 | Department works to externally accredited ISO quality system | 100 | 0 | 100 | 100 | 100 | 0 | 100 | 35 | 0 |
| S2 | Customer satisfaction surveys done regularly | 100 | 0 | 100 | 100 | 100 | 0 | 100 | 35 | 50 |
| S3 | System to prioritise work requests | 80 | 50 | 65 | 100 | 100 | 100 | 100 | 75 | 100 |
| S4 | System to ensure medical devices have regular safety checks | 95 | 100 | 85 | 100 | 60 | 75 | 20 | 80 | 50 |
| S5 | System to ensure medical devices have planned maintenance | 80 | 70 | 85 | 50 | 60 | 75 | 20 | 80 | 50 |
| S6 | Undertake equipment training for medical staff | 80 | 20 | 20 | 0 | 10 | 0 | 0 | 50 | 40 |
| | Overall % score for systems indicators | 89 | 40 | 76 | 75 | 72 | 42 | 57 | 59 | 48 |
| | **Processes** | | | | | | | | | |
| P1 | % of emergency calls responded to within 1 day | 100 | 100 | 100 | 100 | N/A | N/A | 88 | N/A | 5 |
| P2 | % of jobs done in 2 days | 36.6 | 81 | 46.9 | 38.2 | 65.7 | 70 | 46.2 | N/A | 61.8 |
| P3 | % of jobs done in 30 days | 97.8 | 96 | 88.1 | 100 | 92.3 | 100 | 80.7 | N/A | 99.7 |
| P4 | Number of medical devices/Number of beds | 10 | 5 | 8.7 | 10 | 9.5 | 5.7 | 15.1 | 8.4 | 5.4 |
| P5 | Number of medical devices/Number of in house staff | 947 | 415 | 655 | 1651 | 1033 | 800 | 1140 | 1105 | 926 |
| P6 | Cost of contracts/ Number of medical devices on contract | 675 | 862 | 702 | 482 | N/A | 485 | 537 | 1002 | 41.8 |
| P7 | Cost of in-house maintenance as % of total maintenance cost | 51.3 | 36.3 | 33.2 | 41.4 | 16.3 | N/A | 51.5 | 46.5 | 42.6 |
| P8 | Cost of in-house maintenance as % of inventory list price | 3.2 | 1.7 | 1.4 | 1.4 | N/A | 3.24 | 2.6 | N/A | 2.3 |
| P9 | Cost of in-house staff/Number of medical devices maintained | 29.1 | 55.6 | 58.2 | 30.7 | 28 | 25 | 27.7 | 26.7 | 56.6 |
| | **Outcomes** | | | | | | | | | |
| O1 | Incidents/100 Infusion Pumps | 0 | 0 | 1.7 | 0 | 0 | 0 | 4.9 | N/A | 5.9 |
| O2 | Breakdowns/100 Infusion Pumps | 48.4 | 39.5 | 120 | 64 | 71.5 | 67.2 | 94.2 | 45.1 | 23.5 |
| O3 | User errors/100 Infusion Pumps | 4.2 | 1.6 | 13.2 | 18 | 8.5 | 22.4 | 7.8 | N/A | 23.5 |
| O4 | Diversity/100 Infusion Pumps | 1.4 | 6.3 | 1.1 | 4.2 | 4.6 | 4.5 | 2.6 | 9 | 65 |
| O5 | % of staff trained on Infusion Pumps | 75 | N/A | 10 | N/A | 78.7 | N/A | 50 | N/A | 65 |

encourage repair but not investigation of the underlying cause. Peer assessment would be needed to identify whether an event could have been prevented by better quality servicing or user training. This is an illustration of how outcome (effectiveness) is generally more subjective and difficult to quantify than an activity (or process). Performance indicators can also be ambiguous, in both definition and interpretation. A measure such as *equipment downtime* usually refers to how long an individual item is unusable, but supplying a backup unit may mean the clinical service it supports is only out of action for a short period.

### 14.5.3 Selecting and Using Performance Indicators

Appropriate indicators vary depending on what management is trying to use them for. All performance indicators contain hidden assumptions. For example, manufacturers may quote a guaranteed downtime percentage that includes hours when equipment is not needed, whereas clinical users consider that the actual time equipment is out of action during working hours gives a measure more closely related to service disruption. Statistical measures can be particularly difficult to interpret, so *mean repair time* is informative when comparing similar equipment but of limited value across a mixed stock. Process indicators tend to change less frequently than those devised to support a particular improvement, and usually monitor key factors affecting output quality. Outcome measures are more likely to monitor the effectiveness of initiatives to improve the service.

Reasons for poor performance should be investigated. For example, a drop of 10% in 'monthly number of routine maintenance hours logged by workshop staff' could be due to slipping internal standards or alternative external demands such as a need to train a cohort of new users. Steps to improve include modifying processes and procedures, providing more resources or modifying demand. If an indicator is modified to make it more useful, it is also helpful to maintain a comparison with the old one for long enough to check that the change has been advantageous. Process targets should be set and interpreted in the light of local circumstances, as detailed practice often varies, whilst outcome indicators are more directly comparable between services if similar definitions are used.

Cost measures are very sensitive to what a service includes or omits. Annual maintenance cost per item varies widely across a typical hospital, so including a large number of low specification items can drastically reduce the apparent cost of maintenance, as can a re-apportionment of overheads. Detailed examination of costs is more likely therefore to highlight where action to reduce costs will be most effective. For example, the major cost component of infusion procedures is consumables, with device depreciation and maintenance counting for a small proportion of whole life cost. Negotiating consumable prices or combined purchase, maintenance and consumable packages is likely to yield better returns than attempting to squeeze more efficiency from the maintenance operation. Similarly a lifetime approach to costing may help to avoid the situation where a technician spends inordinate amounts of time repairing a piece of equipment because clinical staff do not have funds to replace it. A sense of perspective is needed, for example, so that time spent finding the cheapest supplier for a common component to save a small amount does not detract from urgently repairing a device for theatres to avoid a far greater clinical and financial loss from cancelled surgery. The clinical engineer is responsible for balancing cost and urgency and for justifying this to finance and clinical staff.

## 14.6 Benchmarking in Clinical Engineering

Benchmarking compares the activities and performance of similar services at the behest of service providers themselves. Internally driven rather than externally imposed, it aims to identify good practice and share experience between organisations and so indirectly helps participants achieve compliance with external standards. It is defined as, *'A continuous, systematic process for evaluation of the products, services and work processes of organisations that is recognised as representing best practice for the purpose of organisation improvement'* [13,14]. Costs are an important, but not overriding, element. Principal reasons for benchmarking are to learn from others and improve, to rank organisations and to address costs or cost-effectiveness. The detailed derivation of benchmarking measures can vary between organisations, and enquiring into differences is often more illuminating than making a straight comparison. Its results inform debate when weighing cost against other performance measures and help guide evidence-based decisions which combine quality with value for money.

No engineer should be entirely comfortable with the assertion that their service is the cheapest available. A service should be high quality and provided at a reasonable cost. A benchmarking approach to address both cost and quality might include the following steps:

- Establish a set of local quality measures for the particular aspect of the service being benchmarked, based on a balanced approach but excluding costs.
- Set up a model of the staffing structure and equipment base required to deliver the service to that level of quality, using professional and other guidelines.
- Cost this model by assigning a grading to each post and a realistic budget to equipment and contract service costs.

The resultant cost model can then be compared with similar services. Departments should consider where and why they differ from others and develop their service towards achieving performance as good as or better than that of the best ranked provider, once detailed differences are taken into account. Being clear about costs and methods of overhead allocation is crucial to enable valid comparisons to be made. This approach reflects the main value of costing in benchmarking, which is as an internal tool to help plan and staff activities appropriately. It starts by specifying a quality service, as defined by outputs or outcomes, and then assigns costs to this service rather than focusing on cost as the starting point.

Benchmarking is widespread in equipment management services, most often in repair and maintenance. This is driven by the constant scrutiny of such services for economies between in-house and external provision. Table 14.2 shows the results of annual benchmarking from a local network of departments in England, addressing structural, process and outcome measures. Apart from being difficult to compare directly, the differences between departments on certain measures are greater than would be expected from efficiency variation alone, with, for example, a fourfold variation in the number of medical devices per member of in-house staff. These difference can be explained partly by looking into the detail of what devices are covered and how the individual hospitals operate, leaving a residual variation which indicates where efficiency improvements can be made. The scheme has been the focus of targeted improvements by individual clinical engineering departments and has encouraged joint working on specific initiatives to improve services. Another scheme in the United Kingdom is Clinical Engineering Benchmarking Clubs, affiliated to the UK National Performance Advisory Group [15]. Benchmarking activities include mutual audit; developing common policies on topics such as acceptance procedures for novel equipment, risk assessment testing requirements and minimum maintenance levels for specific equipment types; agreeing equipment classes for database entry; sharing experience of equipment performance and reliability, hazards and incidents; and devising common performance indicators.

In Canada, standards for measurement and evaluation of clinical engineering services were produced in 1998. To put these standards into practice, the Canadian Medical and Biological Engineering Society [16] created a peer review to enhance the sharing of ideas. Peer review looks at the breadth and quality of an organisation's clinical engineering services via a collaborative, impartial and thorough assessment performed by professionals with extensive experience in clinical engineering. Surveyors work alongside the service, often learning as much from the experience as the site being surveyed. Both UK and Canadian systems use peer review and constructive cooperation to drive improvements. An alternative option is to employ external consultants to assess a service or advise on the use of quality systems and performance indicators to improve it.

## 14.7 Audit

Audit is essential to achieving and maintaining quality. At its simplest level, an audit compares current to intended practice and reports on the difference. Audit therefore differs from inspection, which focuses on checking whether processes and their outputs operate within agreed tolerances. For example, an auditor might follow an individual through a process whilst checking for any deviation from relevant procedures, looking at forms and records to check that they are completed correctly. This approach is typical of internal audits and has value in confirming whether a process is working as intended, but does not capture the full power of what an audit can achieve. A more sophisticated audit goes beyond compliance to examine an organisation's intent, as expressed in its strategy and objectives, and its effectiveness in achieving them. The auditor systematically reviews activities to consider the level of quality being delivered, checks how risks are dealt with, and challenges whether existing processes can be improved. A suitably skilled and experienced auditor can usually suggest improvements in the way things are done.

An auditor needs to have at least some degree of independence and not carry direct responsibility for the function or area being audited, something difficult to achieve in a small organisation. This is where external audit brings benefits, particularly when carried out by an experienced colleague from elsewhere. In outline, a typical audit process or inspection visit consists of the following:

- *Planning and preparation*: An auditor will agree to the scope of an audit in advance. For certification, they will review the whole operation over one or more visits. This notice gives the service time to brief staff, review documentation, check if procedures are adhered to and update them where necessary, and get everything ready for the auditor to access.

- *Opening meeting*: The auditor meets with a senior management representative to explain the purpose of the audit and outline the audit timescale.

- *Audit*: This may consist of visits to various areas, interviews with staff and examination of records/databases, usually in the company of someone familiar with the operation. Staff awareness of operational procedures, and their compliance with them, will be checked.

- *Closing meeting*: The auditor will meet with senior management to share audit findings and clarify any queries or non-compliances that might easily be resolved.

- *Audit report*: The auditor writes a report setting out recommendations for improvement or, in the case of formal audits, any requirements for continued certification or accreditation. Once the report has been agreed by both parties, it can then be passed to interested bodies. The service will need to put right any failings quickly.

- *Follow-up*: The auditor will follow up any corrective actions to check if they have been completed, either at the next regular audit visit or sooner if a major problem is found.

Inexperienced auditors may fail to reflect a true picture by distorting the value of particular indicators or insisting on inappropriate remedies. They may concentrate on box ticking without appreciating the relative importance of different requirements, an approach which favours process measures rather than outcomes. More subtly, they may fail to pick up weaknesses at the boundaries between services. A service provider can never assume that a favourable audit result means there are no further improvements to make.

Audit can also be used to assess how well scientific and technical procedures are being followed, and whether they give acceptable results. This type of audit may be carried out by professional bodies under national schemes, for example, cross-checks between radiotherapy departments for dose delivery, and quality assurance schemes for the accuracy of point of care testing systems. It typically looks at outputs, so tests a complete system, which is where it differs to detailed quality control. Audit can be built into commissioning arrangements for health care, for example, in mammography screening in the United Kingdom. It can overlap with benchmarking.

---

## 14.8 Summary

> I prefer a small number of accurate, worthwhile bits of information which people trust and can compare one year to another.
>
> **Professor Sir Brian Jarman [17]**

In this chapter, we have identified reasons why an organisation might monitor its performance and briefly touched on the importance of embedding monitoring in a quality management system. We described the development of performance indicators in general terms and looked at their application to medical equipment management. We discussed some pitfalls in designing and interpreting them and considered their use in practice. We then reviewed how benchmarking can help organisations improve their performance and finally considered how audit can be approached, what happens during an audit and how groups of similar organisations can band together to help each other with mutual audit and benchmarking activities.

# References

1. Spurgeon, P. and Barwell, F. *Overall Performance Measures for Acute NHS Trust Hospitals: Final Report*. Health Services Management Centre, University of Birmingham, Birmingham, U.K., 2000.
2. BS 4778–2:1991. *Quality Vocabulary: Quality Concepts and Related Definitions*. British Standards Institution, London, U.K., 1991.
3. Bittel, L.R. *Management by Exception: Systematizing and Simplifying the Managerial Job*. McGraw-Hill, New York, 1964.
4. Care Quality Commission. *Summary of Regulations, Outcomes and Judgement Framework*. CQC, London, U.K., 2010.
5. Medicine and Healthcare Products Regulatory Authority. DB2006(5) – *Managing Medical Devices, Guidance for Healthcare and Social Services Organisations*. MHRA, London, U.K., 2006.
6. NHS Litigation Authority. The NHS Litigation Authority. http://www.nhsla.com/ (accessed on August 27, 2013).
7. Donabedian, A. Methods for deriving criteria for assessing the quality of medical care. *Med. Care Rev.*, 37(7), 653–698, 1980.
8. Donabedian, A. The quality of care: How can it be assessed? *J. Am. Med. Assoc.*, 260(12), 1743–1748, 1988.
9. Shewhart, W.A. *Economic Control of Quality of Manufactured Product*. Van Nostrand, New York, 1931.
10. Morton, A.P. The use of statistical process control methods in monitoring clinical performance. *Int. J. Qual. Health Care*, 15(4), 361–362, 2003.
11. Center for Development Information and Evaluation. *Performance Monitoring and Evaluation Tips*. US Agency for International Development, Washington, DC, 1996.
12. Pencheon, D. *The Good Indicators Guide: Understanding How to Use and Choose Indicators*. NHS Institute for Innovation and Improvement, Great Britain, 2008.
13. Spendolini, M.J. *The Benchmarking Book*, 2nd edn. AMACOM, American Management Association, New York, 2003.
14. Camp, R.C. *Business Process Benchmarking: Finding and Implementing Best Practices*. Vision Books, New Delhi, India, 2007.
15. U.K. National Performance Advisory Group, Best Value and Benchmarking Groups. http://www.npag.org.uk/ (accessed on August 27, 2013).
16. The Canadian Medical and Biological Engineering Society. http://www.cmbes.ca/ (accessed on August 27, 2013).
17. McLellan, A. Knight of data: Professor Sir Brian Jarman. *Health Serv. J.*, 113(5865), 20–21, 2003.

# Appendix A: Practical Issues in Running an In-House Clinical Engineering Service

## A.1 Introduction

Clinical engineers take responsibility for running all or part of an in-house equipment management service. They can also be responsible for overseeing functions carried out by other parts of their organisation or external service providers. No two services are exactly alike when it comes to what they provide, the range of equipment managed and mix of in-house and contracted out work. In this appendix, we consider how a clinical engineering service can be set up and resourced and comment on some of the many practical issues involved.

## A.2 Scope and Size of the Service

In many hospitals, clinical engineering is the service providing most day-to-day technical management of medical equipment and providing expert advice on its management. This includes advising on procurement, drawing up technical specifications, conducting evaluations and clinical trials, acceptance testing, training, managing and maintaining the equipment inventory, routine and breakdown maintenance including service contract management, condemning and disposal, research and research support, adverse incident handling and investigation and clinical user support. Senior clinical engineers may also have strategic and executive roles in their organisation's overall management structure, Chapter 4.

Clinical engineering equipment management services in the United Kingdom arose, without central direction, in the 1970s and 1980s in response to a need to maintain and support the ever-increasing amount of electronic equipment in hospitals. These roles were picked up by both medical physics departments and estate-based maintenance departments. Early motivations were to reduce costs and equipment downtime, and the technical expertise to repair devices was already available both from scientific departments that had devised and constructed novel devices and from hospital estates departments that already had an electrical repair and maintenance function. At that time, the true costs of maintenance were not always understood, and risk control and regulation was more rudimentary. The concept of equipment management grew up alongside in-house maintenance from the realisation that proactive management of equipment was as important to its safe and reliable operation as maintenance and repair. Growing awareness of the true costs of maintenance and support, and understanding where savings could be made and where true risks lay, led to a greater emphasis on the importance of user training and support and a greater contribution to procurement and contract management. By 2000, active medical equipment management was a requirement of governments, insurers and risk managers. Similar patterns have appeared in other countries as the discipline has developed and evolved, including use of the term 'health technology management' to describe this activity.

Published data on the size, structure and functions of clinical engineering services are rare [1,2]. Regardless of size, however, responsibility for all aspects of equipment management can rest within a single service. A small department might consist of a few senior engineers monitoring outsourced services or one or two technicians providing basic troubleshooting and routine maintenance, perhaps as an outpost of a service spread over several sites or in a small general hospital. At the other end of the scale, tens of staff may deliver a comprehensive service providing all aspects of medical equipment management and be involved in wider professional issues such as teaching and device development.

The types of equipment directly managed and maintained in-house vary with the size and ambition of the service. Most departments support portable equipment such as syringe drivers and vital signs monitors, a number perform direct maintenance on higher-risk devices such as anaesthetic machines and ventilators, and those with the requisite skills tackle items such as linear accelerators and x-ray sets. All these functions will have been developed according to the principles laid out earlier in Chapters 8 and 9.

## A.3 Organisational Structure

Figure A.1 contains a schematic structure for an idealised clinical engineering service of about 20 people which performs all common equipment management functions. The scope of what teams actually cover depends on their size. In a smaller department, quality manager and health and safety roles are part-time responsibilities for a senior technician or engineer. In many organisations, some functions are carried out by other departments, so clinical measurement might be a separate section of an umbrella medical physics department, or support in intensive care be provided out by dedicated

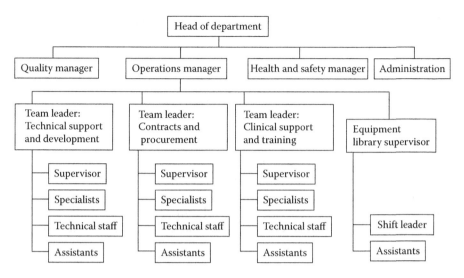

**FIGURE A.1**
Possible organisational structure of a large clinical engineering department. Actual teams will vary in the scope and content of what they cover.

technical staff. Management and administration are present in all services, regardless of size and scope.

## A.3.1 Management Team

The senior management team usually comprises the head of department, senior administrator and internal team leaders. The head of department faces outwards and inwards, undertaking committee work, providing expert advice within the organisation on equipment management issues and resolving problems with the service. Likewise, section heads are part of senior management and also look after the teams they head. The senior management team is responsible for annual business planning including any expansion or contraction of service, budgeting, staff, space and identifying risks. They help to develop policy and highlight any difficulties in its implementation, helping to constantly improve operational procedures in collaboration with clinical users. The management team also sets policy and direction for the quality system and monitors service effectiveness and performance.

## A.3.2 Administration

Considerable paperwork is required to employ staff and use equipment in any service. Staff health and safety, training, sickness management and payroll queries are among many functions requiring monitoring and records. Equipment and spares need to be ordered and items sent off, which generates considerable administrative work chasing orders and payments and dealing with queries. Budget monitoring and control are based on detailed records of spending and monthly review. The closer such administration is to the clinical engineering service, the less the chance of errors due to miscommunication or lack of understanding. Administrators may also provide the initial contact point to log and pass on service requests.

### A.3.3 Procurement

Procurement may be managed by the organisation's central procurement or supplies department or by administration staff if carried out entirely within the clinical engineering service. In the latter model, central supplies/procurement will want to retain an overview and will need, for example, to authorise orders even though the content is generated by clinical engineering.

### A.3.4 Quality

Many departments run under an accredited quality system, which needs a quality manager to develop and oversee it. This role takes substantial time, usually at least half a whole time equivalent in a sizeable department. One or more internal auditors must also contribute and should be independent of the processes they audit. Management of a quality system includes controlling records and procedural documents, and storing them in defined physical or electronic locations. The quality manager keeps the system in auditable shape and progresses the resolution of nonconformities and actions relating to the quality system.

### A.3.5 Training

Large hospitals are likely to have at least one full-time medical device trainer, devoted to end users, with additional input from other members of the technical support team. Trainers are chosen for their good communication skills and practical knowledge and frequently have nursing or technical backgrounds. In addition to training clinical staff to use medical devices, the department may train its own technical and scientific staff and also provide work experience and practical training in clinical engineering.

### A.3.6 Clinical Support

Depending on its size, location and history, a clinical engineering department may include front-line clinical services such as intensive care technicians who provide daytime and perhaps on-call cover. They set up, troubleshoot, advise and train on equipment operation within the intensive care units. They may be responsible for managing consumables and for equipment provision and availability, together with routine calibration and user maintenance, and can manage equipping projects and choice of maintenance arrangements. The service may also be responsible for renal technicians, who support the smooth running of dialysis machines in patients' homes and in hospital. Their work includes installation and maintenance of dialysis machines, patient training, troubleshooting and monitoring water quality. Other technicians who support physiological testing may be closely associated with the clinical engineering service, usually within a medical physics department, performing quality assurance tests and calibration on diagnostic equipment.

### A.3.7 Technical Support

This team is usually the largest single group in a service. Its functions are numerous and include acceptance testing, maintenance, repair, disposal, project work and possibly development work. Technical contributions to the work of the department include procurement, quality and setting policy. Direct maintenance can be a large proportion of the work in some departments and a small part in others (Figure A.2). The issues behind this are discussed in detail in Chapters 8 and 9.

### A.3.8 Health and Safety Representative

The head of service is ultimately responsible for health and safety within their department. Time must be spent on risk assessments and on carrying out any remedial actions. In a large department, an individual could support this as a part-time activity.

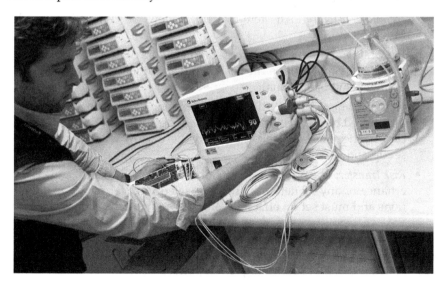

**FIGURE A.2**
Clinical engineer functionally testing medical equipment. (Reproduced by permission of the Institute of Physics and Engineering in Medicine, Copyright IPEM.)

### A.3.9 Equipment Library

Many departments run an equipment library with the help of one or more full-time members of the staff. Typical functions include storage, charging, inspection testing and cleaning of fleets of portable devices such as syringe drivers and vital signs monitors and delivery to and collection from clinical areas.

### A.3.10 IT Specialists

Specific IT expertise is needed in clinical engineering where medical devices are linked to IT networks. This is both to troubleshoot problems and manage the associated risks, which include device–network interfaces, data and information integrity and confidentiality and security. The set up, running and trouble shooting of IT and device networks, including support for both hardware and software, can be difficult because of uncertainty over who is responsible for what, and this needs to be made very clear for each system.

Active cooperation with the IT department and with external suppliers is an essential part of running these installations successfully. Some of the skills required include basic knowledge of networks and systems and specific awareness of the characteristics and support requirements of each medical device system. Issues to manage for new developments include the following:

- *Clinical viability*: Assessing potential solutions in the light of existing and future clinical practice.
- *Technical viability*: An item-by-item review of connectivity and functionality.
- *Practicality*: Will the proposed solution work with existing infrastructure? What will be the cost of any additional investment, both financial and operational?
- *Implementation*: How will the development be funded? Will it be introduced by area, clinical discipline or operational function? What contingency and roll back options are there? How will the project be set up and resourced?
- *Risk transfer*: The organisation retains responsibility for overall governance of any IT network and associated medical device installations and must set up effective contracts and structures to do this.

Further information on the elements required are contained in the risk management standard IEC80001 [3], which complements ISO 14971 in extending risk management to operating environments which link medical devices to IT systems.

Finally, the medical equipment database manager will need to liaise with IT regarding any interface with or operation over the organisation's IT network. This will include any functionality which is accessible to clinical departments, such as equipment library modules, where clinical departments are able to access inventory data or book equipment deliveries.

## A.4 Communications

An in-house service delivers much of its value through long-term relationships with clinical users, for which effective communication is vital. Two areas are particularly important:

### A.4.1 Contact Points

How do users initially contact the service? A single point of contact makes it easier to log requests for assistance and track their progress, making it easier to pick up a task if an individual is absent unexpectedly. However, in practice, there is generally more than one method of contact – usually physical visitors are met by a receptionist with remote enquiries handled through a common telephone, bleep or mobile phone number. Engineers may also carry out routine visits and ad hoc inspections to look for any problems or pickup concerns from clinical staff.

### A.4.2 Communications

Another challenge is to maintain continuity where personnel may change from day to day, particularly with shift-based systems. Computer-based records relating to progress with each task are invaluable, and regular meetings improve communication and service delivery. Table A.1 shows a typical meeting schedule for a medium-sized service.

## A.5 Staff

Engineers and technologists working in clinical engineering departments should have a high level of skills, experience and professionalism, Chapter 7. Routine maintenance of medical devices can be carried out by technicians with lower skill levels, but nothing can substitute for experience and technical and problem-solving skills when it comes to fixing a complex fault safely under pressure. Giving effective advice relies on a wide knowledge of medical procedures, applications and clinical context. This is true of many challenges in equipment management including equipment planning and procurement, development of maintenance policy, user support

**TABLE A.1**

Schedule of Meetings for a Hypothetical Clinical Engineering Department

| Meeting | Schedule | Typical Agenda Items | Attendees |
|---|---|---|---|
| Annual management review of quality system | Annual | Quality policy | Quality manager, senior management representative, head of department, section heads |
| Quality system review | Quarterly | Analysis of key performance indicators Target setting Monitoring non-conformance reports and actions | Quality manager, head of department, section heads |
| Department meeting | Monthly | Operational review Project review Change management Health and safety | All |
| Team meetings | Weekly | Operational issues Targets and progress | Relevant team and section heads |
| Section heads | Monthly | Operational issues Project scheduling and progress | Section heads, admin representative, head of department |

and incident investigation. It is in these areas that clinical engineering services contribute to the reduction of cost and risk across the organisation. The following section outlines the staffing mix found in a typical clinical engineering department.

*Engineers* are concerned with top-down policy, service development and strategy alongside the management of risk, safety and service quality. They provide a very high level of scientific and technical skill and are proactive in formulating and implementing service developments including design, modification, trials and research involving medical devices. They will lead and contribute to the investigation of adverse incidents and manage the medical equipment database in collaboration with those responsible for other asset registers.

*Technologists* have the necessary skills and experience to perform all routine technical functions. These include acceptance testing, safety testing, planned preventive maintenance, troubleshooting, fault diagnosis, breakdown repair, resolving user problems and providing technical training. They may design and construct minor development projects. It is advisable to develop specialist expertise in more than one person so that cross cover is available.

*Assistants* in a larger department may have little or no technical training but can still support the department by acting, for example, as equipment library assistants delivering and caring for equipment, maintaining spare stocks and receiving/despatching equipment.

*Other professionals* such as clerical staff provide administrative and help desk support, and, in very large departments, dedicated managers may support management, procurement, training or finance. Training and equipment library functions are often carried out by technologists or engineers but can be performed equally well by nurses.

Optimum skill mix will vary considerably depending on departmental profile and workload. There are likely to be fewer engineers than technologists and more assistants where work is largely routine. Local staffing models are best developed using information from national professional bodies and data from benchmarking with similar and contrasting departments.

## A.6 Space and Physical Layout

The physical layout of a department ideally reflects how it works. Figure A.3 is a functional schematic of an imaginary clinical engineering department, showing how it might be organised into zones with adjacent related functions supporting an efficient flow of people, materials and equipment through the system. Areas include the following:

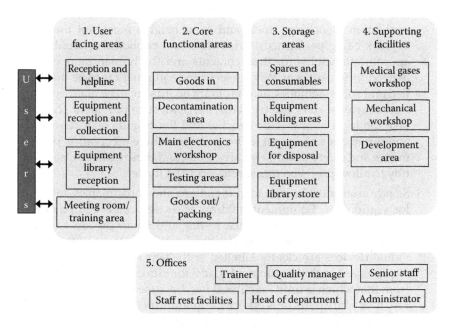

**FIGURE A.3**
Clinical engineering department – schematic functional layout.

1. *User facing areas and reception* need to be staffed at agreed and publicised hours and provide space to meet callers comfortably. It is usually the point to which callers bring small items for repair and from where they collect completed jobs. It is more efficient if this area hosts the department's main telephone contact point, and it should be close both to the main workshop and to the administrative office space. The size of most operations will not require a dedicated receptionist, so this work can be carried out alongside other administrative tasks in lulls between callers and telephone calls. Reception should also be adjacent to goods inwards so it is easy to accept and sign for deliveries of spares, consumables and new equipment.

2. *Core functional areas* include *goods in*, which holds deliveries of equipment, consumables, spares and components prior to unpacking and acceptance checking. This zone may be adjacent to reception or workshops and be managed by administrative staff or assistants. *Goods out* is an area for packing and holding equipment awaiting return to manufacturers, and a *decontamination area* allows for cleaning prior to servicing or testing, where exceptionally this has not been addressed by decontamination procedures elsewhere in the organisation. The *main electronics workshop* is where smaller portable items such as syringe drivers and monitors are serviced and repaired. Health and safety risks should be addressed, with sufficient bench space to use and store test equipment and tools safely, isolated electrical supplies and residual current breakers, and solder fume extraction or good ventilation. Benching, stools and chair heights must suit the kind of work being performed, and shelves and cupboards should be at appropriate heights for their contents and the human manoeuvres needed to access them. Where electronic equipment is to be dismantled and sub-assemblies such as circuit boards are to be handled, protection against static electrical discharge must be provided. This can take the form of anti-static benches and floors and conductive mats and wristbands. Adjacent *testing areas* are where acceptance, performance and safety testing are carried out, and these must be clean and tidy to allow precision measurements to be made.

3. *Storage areas* must be provided for spares and consumables and also for equipment. Equipment holding areas segregate equipment at various stages of the acceptance, testing and repair process to facilitate tracking and should be clearly designated areas where individual devices are clearly labelled with their status. An inviolable requirement is that equipment ready for clinical use is segregated from that which is unfit for clinical use, because the latter is awaiting repair or spare parts or is otherwise quarantined or due for disposal.

4. *Supporting facilities* can include a *medical gases workshop* to facilitate testing and repair of devices that use medical grade gases, including

critical care ventilators and anaesthetic machines. It should have piped medical gases and vacuum and be able to accommodate larger portable items including ultrasound and endoscopy systems. Cylinders of oxygen and air can be used instead of a piped supply, but supply and storage is a problem if there is a high workload. A local air compressor can be used but should deliver gases of the high standards of purity required from a piped system. Medical vacuum, of course, cannot be provided by a cylinder and so piped supplies should be provided to the clinical engineering workshop, unless its distance from the main supply is so great that there are insurmountable installation costs or other problems. A *mechanical workshop* can vary from a single workbench, pillar drill and hand tools to a fully equipped development facility with computer-controlled lathes and milling machines; grinding, welding and sheet metal working facilities; and injection moulding. Even basic facilities can help to sort out mechanical equipment problems if skilled staff are available, on items such as chairs, trolleys and even ventilator parts. A workshop may design and develop custom medical devices, and most demands are not for innovative or complex systems but for local solutions to workaday requirements such as securing equipment to trolleys. However, a workshop may become involved in more complex development projects that form part of student or trainee projects, doctorial research or large-scale funded clinical research. A *development area* supports the design, testing and fabrication of one-off devices for new clinical applications or research use.

5. *Offices* can be open plan, but some closed space is necessary for small private and confidential meetings. A larger *meeting room* can be suitable for training if space is available. *Staff rest facilities* may be provided if the size of the department justifies it.

## A.7 Finance

Wages will be the single highest outlay for a clinical engineering service, accounting for a significant proportion of the department's annual operating budget. In addition, there may be considerable financial responsibility for capital projects and the administration of service contracts. A service may be able to argue for central funding for an organisation-wide standardisation programme, such as the coordinated introduction of a new type of defibrillator or blood glucose meter. The service may be allocated a considerable lump sum to place and administer service contracts on major equipment items such as CT scanners that are not maintained in-house. There will be budget lines for smaller service contracts and ad hoc bills for manufacturer call outs and equipment repairs, and also for spares and consumables, but these are

**TABLE A.2**

Example Cost Centres for a Hypothetical Clinical
Engineering Department Budget

| **Expenditure** | |
| --- | --- |
| Salaries | Scientific and technical |
| | Administrative and clerical |
| Maintenance | Service contracts |
| | Ad hoc repairs |
| | Spares and consumables |
| Equipment | Revenue equipment purchase |
| Departmental costs | Office equipment and stationery |
| | Travel |
| | Training |
| | External advice |
| | Quality System |
| Workshop costs | Tools |
| | Consumables |
| | Test equipment |
| Internal service charges | Telephones, IT support |
| **Income** | |
| External contracts | |
| Cost recovery from maintenance charges | |
| Project recharges | |

often recharged to the end clinical user where budgets are devolved to ward or department level. Training and travel funding is required – these budget lines are often under-funded but, again, must not be forgotten in the budgeting process. In practice, bursaries from within the organisations staff development fund or from professional organisations may provide a substantial part of this funding. Table A.2 shows a number of possible cost centres in a typical clinical engineering budget.

# References

1. Frize, M. Results of an international survey of clinical engineering departments. *Med. Biol. Eng. Comput.*, 28(2), 160–165, 1990.
2. Frize, M., Cao, X., and Roy, I. Survey of clinical engineering in developing countries and model for technology acquisition and diffusion. *Proceedings of the 2005 IEEE Engineering in Medicine and Biology 27th Annual Conference*, Shanghai, China, 1–4 September 2005.
3. IEC 80001-1. *Application of risk management for IT-networks incorporating medical devices – Part 1: Roles, responsibilities and activities.* International Electrotechnical Commission, Geneva, Switzerland, 2010.

# Appendix B: Electrical Safety for Medical Equipment

## B.1 Introduction: Medical Equipment Is Different

Electrical equipment presents risks in the workplace that are minimised by regulatory standards for design and construction and by safe installation and use, coupled with inspection and maintenance by competent personnel. In addition to these measures, the healthcare professional must take extra precautions when working with electricity in the clinical environment. There are duties of care to protect patients, as well as staff who operate and maintain equipment from the special hazards associated with medical device use.

Medical equipment differs from domestic and industrial equipment in a number of ways. There can be a direct electrical pathway to the patient, as with ECG electrodes. These can be internal in the case of invasive catheters or intra-operative probes. When patients are unconscious or sedated, they cannot react appropriately to shocks or burns. The clinical environment contains conducting fluids such as saline and blood. Flammable or explosive vapours such as oxygen, alcohols and anaesthetic vapours compound the hazards of sparks and overheating. Electrical standards for medical equipment are more onerous than those for domestic or office equipment, requiring tighter control of leakage currents, for example, with limits that are specified under both normal operation and a range of single fault conditions. Special safety considerations in the clinical environment do not just apply to medical devices but to any electrical device used near the patient, or connected to one that is. They also apply to the electrical infrastructure in critical clinical areas, such as operating theatres, where special regulations exist that are more demanding of installation safety than in standard domestic or industrial circumstances.

## B.2 General Electrical Hazards

Adverse effects of electricity that could be experienced by patients or staff range from mild discomfort to death by ventricular fibrillation (VF) or respiratory muscle paralysis. They include injuries due to shock, burns or

**TABLE B.1**

Hazards Associated with Various Categories of Electrical Exposure

| Electrical Category | Common Sources in Medical Equipment | Hazard(s) |
|---|---|---|
| Static charge | Insulated dry environments, carpets, floor surfaces | Shock leading to disorientation<br>Fire or explosion in the presence of flammable or explosive gases and vapours<br>Damage to sensitive electronic circuits during workshop repair |
| Mains electricity | Universal | Shock<br>Respiratory paralysis<br>Cardiac arrest<br>Burns<br>Fire |
| AC leakage | Electrodes, enclosures | Cardiac arrest |
| Radiofrequency power | Diathermy, physiotherapy | Burns<br>Interference to equipment |
| High voltage | CRT monitors, radiotherapy, x-ray | Shock<br>Death<br>Burns<br>Fire |
| Electromagnetic radiation | Radio communication equipment | Interference to medical equipment; function of life-critical devices affected, diagnostic quality affected |
| DC | Low voltage on applied electrodes | Sores, skin damage |

secondary effects of shock such as falls and sudden muscular reflexes when working with hazardous machinery or dangerous materials. Electrical faults are a common cause of fires, and in a hospital, these produce added risks caused by disruption to routine clinical services. Table B.1 summarises these risks, which are explained in more detail in the succeeding text.

The physiological effects of electricity are outlined in Table B.2. When a source of electric current at low level is connected to a human, nothing will be felt below a current threshold of around 1 mA (see Section B.8 for an explanation of units). Anything above this level will produce an electric shock which will have more severe effects on the nervous system as the current magnitude increases. At lower currents, the reflex action will be hopefully to pull away. At higher currents, muscle spasm and disorientation may take place, and a current entering the body of around 100 mA is capable of bringing death by VF or respiratory arrest if it flows through sensitive organs. The internal current through the heart to cause VF is around 0.1 mA, and there are obvious hazards if such a small current is applied directly through the heart as, for example, during surgery. If the shock victim is fortunate to have assistance, resuscitation may be

**TABLE B.2**

Physiological Effects of Electric Current

| Current | Effect |
|---|---|
| External, 1 mA | Current perceptible |
| External, 100 mA | VF, respiratory arrest |
| External, 200 mA | Serious burns |
| Internal, 100 μA | VF |

possible. However, at still higher currents (200 mA and above), additional effects such as major burns can cause serious complications.

*Static electricity*: Is usually produced by friction in an environment, such as a person's shoes rubbing against a carpet when walking in a dry office. Electric shocks delivered when an earthed metal object is touched are usually harmless – although the electric potential differences involved can be tens of thousands of volts, the stored charge and consequent electric current are too small to constitute a danger. However, the effects of static electric shock are unpleasant and can have unfortunate consequences – a member of staff may drop an expensive item, for example, or worse still make an error during a patient treatment. In a repair workshop, the high potential of static electricity can destroy some electronic devices (integrated circuits) causing damage invisible to the eye but requiring the replacement of thousands of pounds worth of components. In an environment containing explosive or highly flammable gases such as operating theatres or laboratories, a spark generated by static electricity can be highly dangerous. The effects of static electricity can be alleviated by the use of anti-static materials and devices in clinical and workshop areas, including specialised conductive floor coverings and work-top surfaces across which charge cannot build up. In the workshop environment, technologists need to take anti-static precautions such as repairing electronic equipment on conductive mats to which they are connected by wrist straps. Gas pipes and flexible hoses may be made conducting to prevent the build up of static electricity, and explosions and fires have been caused by replacing conducting (anti-static) oxygen hoses with non-conductive tubes.

*Alternating current and direct current*: Direct current (DC) flows in one direction only, whereas alternating current (AC) reverses flow rapidly – 50 times per second (Hz) in the United Kingdom and 60 Hz in the United States. AC is a more efficient means of electrical power transmission, so mains power supply almost universally uses this form of electricity. AC can cause muscle paralysis, rendering ineffective the reflex action to pull away from a source of electricity when a current is felt. When attaching electrodes to patients any stray DC, even at low voltages, inadvertently applied to skin electrodes for any length of time can cause sores and other skin damage due to electrolysis, whereas a comparable AC current does not cause these effects. The physiological effects of a small AC current become less hazardous at increasing frequency, hence higher *leakage currents* are allowed (see succeeding text).

*Health and safety hazards*: The effects of electric shock, reflex muscle action, disorientation or unconsciousness, can lead to secondary injuries more serious than those caused by electricity itself. Reflex action may cause an individual to collide with hard surfaces or tear their skin on sharp ones. In critical situations such as operating theatres, even a minor shock to a member of the staff can cause distraction, leading to an error or accident (see Chapter 10). Operating faulty equipment can cause further damage to the equipment itself, which may undermine a clinical service. Equipment can become damaged and unsafe by constant movement and handling by multiple users who treat it with various degrees of care.

*Fire or toxic gases* can be caused by overheating of electrical cables or when failed components in a device catch fire. Many devices are fitted with internal cooling fans, and if these fail, equipment temperatures may reach dangerous levels. Motors in particular are designed to dissipate heat produced during normal running, but if they become stalled or seized, very high current flows can cause severe overheating with emission of flammable gases from insulating materials and greases which can then be ignited by the high temperatures produced.

## B.3 Risk Management of Electrical Hazards

There are three ways to address electrical risks in healthcare. Firstly, design and construction of equipment must minimise danger and ensure all foreseeable operational risks are taken into account. This requires compliance with standards and guidelines and conformity with the relevant regulations. Secondly, equipment must be used safely. This requires adequate user training, the provision of user instructions and appropriate checks before use. Thirdly, risk-assessed maintenance and inspection programmes will check for long-term deterioration and the need for repair or replacement.

Electrical safety in a healthcare organisation can be promoted at all stages of the equipment life cycle, from procurement to disposal. All devices obtained for routine clinical services must comply with appropriate medical device safety standards, and any experimental equipment must be carefully managed. The underlying electrical power infrastructure must be safely installed by a competent person and be inspected and tested at appropriate intervals, and specific national requirements usually apply to hospital and healthcare power supplies and to fixed medical electrical installations. Medical equipment acceptance testing includes electrical inspection and tests where appropriate, based on international standards for the electrical safety of medical equipment [1], for the routine electrical testing of medical electrical equipment [2], and in the United Kingdom, the testing of portable electrical appliances [3].

Training in safe equipment use is the first essential, including background training in basic electrical safety issues, what to look out for and how to report any damage or malfunction. Equipment should be inspected before each use and be maintained, inspected and tested at defined intervals by competent persons (see Chapter 8). It should be visibly labelled with a fitness for use label and not used if beyond an inspection due date. Even when equipment is out of use, condemned or awaiting disposal, it should be stored in a safe condition, with precautions taken to prevent inadvertent use or abuse (see Chapter 12). In most countries, it is a legal requirement to maintain working devices in good repair and in a safe condition, and this requires day-to-day inspection and alertness on the part of equipment users.

## B.4 Basic Principles of Electrical Safety in the Design and Construction of Equipment

General principles of electrical safety apply to all electrical devices, with further refinements that meet the special conditions of medical use. The general principles and standards of construction and protection of devices include protective earth connections or double insulation and automatic supply disconnection in the event of overcurrent or overheating conditions. The specific principles and standards of construction and protection of medical devices include additional isolation of patient-applied parts, restricted leakage currents if a single fault occurs and immunity to a minimum level of electrical interference and induced currents. In this section, we outline basic electrical protection against fault currents by fuses and earth connections (grounding). In the next section, we consider special requirements for medical devices, particularly for protection against leakage currents.

*Fuses and current trips*: The electric current entering a device can be interrupted if it becomes too large. Fuses and overcurrent trips are the main protection against fire, removing the source if too heavy a current is passed. Direct sensing of overheating by temperature sensors is employed for some motors and electrical transformers where a fault current not high enough to blow a fuse can still cause dangerous levels of heating. Thermal fuses are designed to blow at a specified current and a specified temperature. They thus cut off systems when the temperature around them rises unacceptably for any reason, as well as protecting against excessive currents. It is crucial to the prevention of electrical fires that protective devices are not circumvented, and if they require replacement, an exact equivalent must be used. Fuses contain thin wires that melt when the current exceeds a predetermined level for a minimum time period, and each application has a specified fuse current rating. In the United Kingdom, mains plugs are fitted with internal fuses that should be matched to the appliance to which they are fitted. If a device fitted with a 5 A-rated mains cable is used with a 13 A fuse in the mains plug, a fault current of 10 A could flow for long periods and could eventually result in overheating and a fire. A residual current device (RCD) protects people against electric shock by switching a circuit off in a period of tens of milliseconds when a low current (typically 30 mA) is detected leaking to earth, potentially through a person. It detects any imbalance in the currents flowing in the live and neutral mains conductors. RCDs, by the same mechanism, reduce the risk of fire by detecting electrical leakage to earth in electrical wiring and accessories. They are used more commonly in laboratories or with portable tools rather than with medical devices. However, their use in workshops where medical equipment is tested or maintained is very widespread, and such forms of protection should be an integral part of workshop design for clinical engineering services.

*Earthing*: Electrical devices require two supply wires to operate – the *live* wire to carry current to the device and a *neutral* connection to complete the electrical circuit back to the generating station. There are two approaches to earthing, class 1 and class 2. Class 1 equipment has a protective *earth* wire linked to the equipment case, which greatly reduces any fault currents caused by loose internal live wires or insulation failure flowing through any person touching the conductive metal case by providing an alternative low-resistance current path to earth. With a low-resistance path along the earth wire, a high current should flow if there is a fault which also performs a secondary protective function by causing the fuses or trip to blow and cut off the unit's mains supply. The earth wire is connected to neutral at the power station to reference the supply voltage to earth (see Figure B.1) and

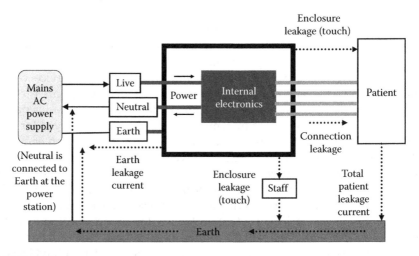

**FIGURE B.1**
Leakage currents for an earthed (class 1) medical device, showing potential flows through patient, staff and the earth connection from the equipment enclosure and from electrical connections to the patient.

is also connected physically to ground at strategic points along its length, such as near where it enters a building, to avoid potential differences building up due to stray and induced currents. Class 2 equipment has no protective earth but relies on robust insulation – in two layers – to separate any fault from a user or patient. Double insulated devices still require fuses or other overcurrent protection, however, to prevent internal faults causing overheating and fire and insulation must be inspected regularly for damage.

## B.5 Special Requirements for Medical Devices

The special dangers presented by medical devices require more subtle safety considerations than for domestic equipment, particularly in the case of leakage currents. These considerations are rarely appreciated by nontechnical users without instruction, and consequently, certain electrical safety precautions may be ignored. It is up to those with technical knowledge to educate end users in the subtleties of medical equipment safety.

The parts of a device in contact with the patient are known as *applied parts* and include ECG electrodes and the like. They form a deliberate direct electrical connection with the patient and include any part of the equipment that intentionally comes into physical contact with the patient internally or externally, such as ultrasound transducers and endoscopes, even though these are not designed to provide electrical contact. Although there may be no actual metallic path between some of these connections and the mains supply, AC can be transmitted through substances normally thought of as insulators by capacitive coupling from the mains supply, and this can be sufficient to cause microshock currents that are hazardous and even fatal under certain circumstances – such as if they end up being applied directly to the heart. AC *leakage currents* can flow into the patient from applied parts such as electrodes. It is also possible for leakage currents to flow between applied electrodes, and these are known as auxiliary currents. Protective earth (class 1 devices) cannot avoid coupling of the AC mains current to the metal enclosure as well as to any applied parts, to create an earth leakage current. This current is mostly conducted harmlessly to earth via the earth wire, with the possibility of a tiny current flowing to ground through anybody touching the case. These currents are known as *enclosure leakage currents* or *touch currents* and are usually very small when the earth wire is functional. However, should the earth wire become disconnected, perhaps by mechanical damage to the mains cable, the earth leakage current can be diverted through the enclosure and flow through anybody who touches the equipment. Electrical testing measures whether enclosure leakage currents are low enough, both during normal circumstances and also under a single fault condition. There are two of these: a disconnected (*open circuit*) earth, as just described, and a disconnect (*open circuit*) neutral. Under normal conditions, the live wire is at the full mains potential and the neutral wire close to zero. If a single fault condition occurs where the neutral wire is disconnected, any internal electronics of the device connected to the live wire will rise to the potential of the mains supply. Hence, although the device will not operate, and the normal mains operating current will not flow, higher than usual internal potentials can cause increased leakage currents.

A further variant of each leakage current test is performed with the mains polarity reversed. Leakage currents depend on the precise electrical design and layout of a device, and usually, the mains connection is designed to be symmetrical, with fuses and switching in both live and neutral wires. The two wires generally follow similar paths within the equipment, with any interference or surge suppression components applied equally to both. In this case, should the mains supply be reversed, due either to incorrect wiring of the mains socket or plug or the ability to insert the mains plug in the socket in either way in some countries, similar earth and enclosure leakages are likely from the mains part as with the mains polarity in the normal direction. However, where the mains part is not symmetrical, leakage currents can be substantially different with reversed polarity. It is thus necessary to test devices under both conditions.

Another possible hazardous condition is if a patient is connected to one device via an applied part and touches a source of current, for example, another piece of equipment with a serious fault. Current can then flow through the patient from the faulty device into the applied part. Medical device standards require that the maximum current into the applied parts under these conditions is limited. The *mains on applied part* test measures this current with the applied part directly connected to the mains live terminal.

One way to increase intrinsic electrical safety in a clinical environment is to remove the earth connection from the patient altogether and leave them electrically *floating*. Thus, operating tables, trolleys and beds may be electrically insulated from their wheels, so that a ground path is not available through the patient. When this is done, a very high resistance path must be provided so that static electricity does not build up with its inherent risk of sparking that can cause equipment damage, shocks to people and explosion hazards. Mains isolation transformers provide additional protection against both gross electric shock and microshock from individual items of equipment, interconnected systems and in whole areas such as operating theatres. They ensure that there is no direct reference to ground potential on either terminal of the secondary winding of the isolation transformer (see Figure B.2).

The primary winding of the isolating transformer is connected to the mains supply, whilst the secondary winding is designed to produce the required voltage and current capability as would be available directly from the mains supply but without any direct connection to it. This means that if either wire of the secondary is accidentally touched by a person standing on conducting ground, the absolute potential of the wire will fall to that of the person,

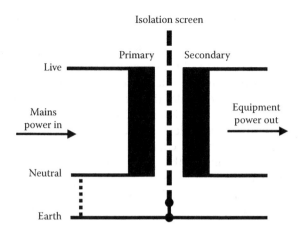

**FIGURE B.2**
Mains isolation transformer protection with floating power output.

and no dangerous current will flow. Note however that the potential difference between the two wires will be maintained, so a person grasping one in each hand is likely to receive as fatal a shock as they would from direct mains contact! Leakage currents from the secondary are limited by the presence of a conducting non-magnetic screen placed between the primary and secondary windings. This screen is earthed and helps both by conducting leakage currents to earth and by considerably reducing the capacitance between the two windings. This is the usual principal of operation of isolating transformers used in mains supplies, to either items of equipment or whole installations in buildings.

Electrical isolation of signal paths between items of equipment can be effected with electro-optical coupling or wireless transmission. Within medical equipment itself, two electrical isolation techniques are commonly used to limit leakage currents through patient-applied parts. The front end electronics of medical devices such as an ECG machine will be isolated from the remainder of the electronics by electromagnetic coupling, used to supply the power to the front end electronics, and electro-optical coupling used to transfer signal and control voltages. Electronic signals from the patient-connected part of a device can be transferred through optical isolators that, by using light rather than electrical wiring to transmit information, can reduce patient leakage currents to very low levels. The optical path within equipment is usually air, but fibre optic cables (made of highly insulating material) can conveniently be used to transmit physiological signals and digital data between devices.

High leakage currents can be generated in non-medical devices such as computers and some industrial equipment by switched mode power supplies. These convert mains power firstly to a DC voltage (by rectification) and then use high frequency switching to produce square wave AC at much higher frequencies than those of the original mains supply. These are passed through an isolation transformer, which can be made much smaller at high frequencies than at mains supply frequencies, and rectified again to produce DC supplies to power various internal components. Because frequencies used are many thousands of cycles (high kHz region), and there can be substantial power transferred, there is a possibility of electromagnetic emissions which must be carefully controlled, otherwise these emissions can find their way to the enclosure and mains cable and create an intrinsically high earth leakage current and interfere with equipment operation.

The earthing of medical equipment can be more complex with class 2 double insulation being further divided into types which designate the degree of electrical isolation of the patient-applied parts. Type B has the least stringent specifications, and this equipment may have applied parts connected to earth potential, are generally not conductive or may have no part that actually contacts the patient. In type BF (*floating*) devices, the applied part is isolated by one of the methods described in the earlier text; type CF equipment has isolation to a more stringent limit, such that direct cardiac connection is possible. Some devices may be constructed to floating specifications, even though they have no direct electrical contact with the patient, if there is a possibility of an electrical path being established by other means. Defibrillator-proof equipment can be of type BF or CF and is able to withstand the high voltages produced by a defibrillator, so need not be disconnected from a patient should emergency resuscitation become necessary.

Electromagnetic emission limitation and immunity are attained by careful design and experiment using circuit design principles, physical component placement, metal shielding and the incorporation of filters into equipment and cables. These measures often rely on critically sensitive and counterintuitive processes, and particular care must be taken during dismantling and maintenance not to inadvertently disturb these arrangements by careless omission or inadequate tightening of screws, for example. Regulatory approval to place a device on the market relies on rigorous testing of the final product, which cannot be modified without retesting. General hazards from electromagnetic interference from mobile phones, radio transmitters, walkietalkies and other sources are discussed in Chapter 3.

## B.6 Electrical Safety at All Stages of the Equipment Life Cycle

### B.6.1 Procurement and Commissioning

At the procurement stage, specification to appropriate standards, such as CE marking, is essential to show that a medical device is fit for its intended purpose. When putting the equipment into service, risk-based decisions should be made on the frequency and nature of routine maintenance, how breakdown maintenance is to be provided, and what inspections and tests are included in this (Chapter 8). The support and training requirements for end users, on both general electrical and equipment-specific safety, as provided by the manufacturer and in-house trainers, should also be considered. Acceptance checks and those on installation and commissioning will include any necessary electrical safety checks and tests.

### B.6.2 Maintaining Safety in Use

Electrical safety when equipment is in service depends on users correctly carrying out pre-use checks. Users must be trained in the use of the particular device and understand basic principles of electrical safety. Some equipment, such as a vital signs monitor, may rely on quick visual inspection, whilst more complex and life critical devices such as anaesthetic machines will have a pre-use protocol and checklist to run through. In a visual check, the user should inspect the device for *in-date* information, conveniently displayed on a sticker bearing the year and month to which the device can be used, colour coded in the same way as a car tax disk that changes from year to year. Users should be trained to report and not to use equipment that is out of date or, as sometimes happens, belongs to a different organisation as revealed by its asset label. A quick visual check of equipment and its environment will reveal any immediate gross hazards such as damage to cables and connectors, fraying of insulation, cracking of plug covers, loose cables at the point of entry to a plug and any damage to the equipment case itself. Staff should be trained not to attempt their own repairs with, for example, adhesive tape that might not be electrically or mechanically effective and may impose an infection risk. They should report any external or internal contamination by, for example, saline solution that could be electrically conductive and present a hazard to the user or equipment.

### B.6.3 Specific Hazards from Some Device Types

Some devices have specific electrical hazards that should be emphasised in user training. Two examples are defibrillators and diathermy. Both are designed to impart significant energy into patients, the first to restart or resynchronise heart rhythm by applying a short burst of a high voltage to the chest and the second to cut through tissue using a radio frequency current at high voltage to vapourise it during surgery. Defibrillators store and apply substantial charge at potentially lethal voltages. Special precautions are necessary when using and working on them. Resuscitation staff should be aware of the danger of shock from inadvertent contact with paddles or patient. Medical equipment with direct connection to patients should be *defibrillator proof* and designed to withstand the high voltages found in defibrillation. Diathermy is prone to causing unintended burns from poor or inadequate return electrode size or positioning. The current that flows through the patient at the site of cutting should be concentrated over a small area – the diathermy cutting tool – where the high density of current causes the required heating. The other contact to the patient, the return electrode, must have a relatively large surface area to avoid unwanted heating. If this contact is poor and effectively restricted to a small surface area, then burns can be caused at the return electrode. Inadvertent return paths can also cause burns to staff or patients.

### B.6.4 Interconnections

With the growth in the number of electrical items in use, there is often a shortage of mains sockets and distribution boards, and extension leads will be in demand at some time or other. The first rule for users should be to avoid these if at all possible – some standards ban their use altogether – and seek the safer solution of extra sockets. Should their use become clinically unavoidable, a type approved for use by the clinical engineering department must be installed which is selected for robust construction, adequate earth continuity and bonding, cable rating and mechanical strength. Where medical equipment systems are assembled, with for example more than one piece of equipment sharing a distribution board or trolley and with interconnections between devices, appropriate standards for the safety of medical equipment systems as well as individual devices must be followed and testing carried out accordingly [4].

### B.6.5 Computer Standards

Computers can compromise medical electrical safety standards by breaching earth leakage limits for medical systems, particularly as they have intrinsically high leakage currents due to surge limiting and interference suppressing capacitors connected between the mains part and earth and their use of switched mode power supplies. Domestic or industrial equipment can similarly breach leakage current requirements, and must not be plugged into distribution boards for medical systems. As well as leakage current increases, some high-current devices such as kettles and heaters might cause overcurrent trips to operate, leaving medical devices without power.

## B.7 Electrical Inspections and Tests

There are two motives for inspecting and testing medical devices. The first is common to all electrical equipment in the workplace and is the legal obligation to maintain it in a safe condition, to avoid danger. In the United Kingdom, this is controlled by the Electricity at Work Regulations 1989 [3] which cover portable appliance testing (PAT). The second motive is the need to maintain special electrical requirements for medical electrical equipment, as specified in IEC standards 60601 [1] and 62353 [2]. Here, we compare and contrast these approaches and consider how their common elements can be used in various inspection and testing protocols for medical and other devices in the clinical environment.

A device that passes testing to medical standards protocols, where test limits are equal to or more stringent than PAT testing limits for all measured quantities, will be compliant automatically with Electricity at Work requirements. Hence, there is no point in performing duplicate PAT tests on medical devices if they are already tested to medical standards. PAT standards are not medical standards but are usually intended for use on equipment outside the patient environment. In general, it is better to leave testing of medical devices to medical device specialists, especially as they may be prone to damage or arbitrary failure from PAT test personnel unfamiliar with medical devices. When non-medical devices are used in the near-patient environment, they must be made safe by electrical isolation. The Institution of Engineering and Technology (IET) Code of Practice [4] recognises the different situations under which tests are carried out, and this division is helpful in considering which standards are applicable to each situation and which tests are relevant to equipment management in a healthcare unit. *Type tests* are carried out at the design, prototype and preproduction stages of device development to ensure conformity with the regulations in every detail. *Production tests* are carried out by manufacturers to verify individual devices or batch samples fall within desired limits. They test representative functions but not those which might cause destruction or weaken the device. For example, high-voltage tests and drop tests will not usually be carried out on completed production devices. Type tests and production tests will both be carried out by manufacturers and not end user institutions. Acceptance tests will be carried out by end user institutions and will be based on risk-assessed protocols as discussed in Chapters 3 and 8 and later in this appendix. *In-service testing* is carried out routinely to determine whether equipment is in a satisfactory condition, usually as part of a risk-assessed protocol. Tests after repair may be performed by manufacturers or service agents, including in-house services, again as part of a risk-assessed protocol.

Of these four categories, healthcare organisations are concerned primarily with acceptance and in-service testing of medical devices. It has been very common to perform in-service electrical safety tests on medical equipment to a protocol specified by the relevant parts of the IEC 60601 standard. These typically include insulation, earth bonding, leakage currents, auxiliary currents and *mains on applied part* tests. Leakage current tests are carried out under normal and single fault conditions, with mains correctly connected, open circuit and reversed (the latter not considered a fault condition but a variant of normal operation). IEC 60601 tests are type tests intended for manufacturers' compliance but have been commonly adopted by clinical engineering departments as test protocols in the absence of a specific testing standard and because commercial suppliers of test equipment have based their devices on these standards. They have also been widely used by manufacturers' service engineers for testing after maintenance. Prior to the introduction of IEC 62353, there was no international standard for acceptance or in-service testing for medical devices. Some national bodies have developed guidance documents for equipment management that include modified versions of some of the IEC 60601 tests, and in general, these have sought to avoid tests potentially hazardous to equipment or testing personnel or to reduce the applied test currents or voltages to less hazardous levels. With the advent of IEC 62353, a standard specifically intended for acceptance and routine testing of medical devices in service, IEC 60601 is expected to play a lesser role. We will not quote all the detailed test limits in the standards here, as they are widely quoted on many Internet sites and are available in the standards themselves and in guidance documents. We only discuss some of the main differences between IEC 62353 and generally accepted IEC 60601 tests. These are the following:

- Records of tests are required.
- Physical inspection is required.
- Lower earth bonding test current (100 mA) with new limits for acceptable resistance.
- No *mains on applied part* test is required.

One important difference between the medical and the portable appliance standards is that for medical standards, leakage currents are measured and safe limits are defined under both normal operating conditions and single fault conditions.

Given the very low number of electrical faults found on delivered equipment, some organisations find that the resources used on full electrical tests are better diverted elsewhere. In the case of class 2 equipment, visual inspection can suffice, and in the case of class 1, visual inspection combined with earth continuity might seem a sensible minimum. If certification of electrical testing can be obtained from the manufacturer for all items of delivered equipment, then acceptance procedures might dispense with electrical acceptance testing entirely and rely on visual inspection and functional testing. It is up to the individual service to decide whether or not to accept a cost, risk and benefit analysis for a given equipment type that indicates this approach is acceptable. Decisions can be supported by collective action at benchmarking groups. Ultimately, however, the need to test at a risk-based interval to comply with the requirements of health and safety law cannot be ignored, and tests must be performed at some stage. In the United Kingdom, the Electricity at Work Regulations 1989 [3] require that all electrical systems, including electrical appliances, are maintained so far as is reasonably practicable to prevent danger. A competent person must inspect the installation regularly in any public building or place that people work. Guidance from the IET and the United Kingdom Health and Safety Executive suggests initial intervals for combined inspection and testing, by a competent person, that range from three months (for construction equipment) to 1 year for inspection and, in many cases, longer periods for testing (certain types of appliance in schools, hotels, offices and shops). This testing is carried out as a routine to determine whether equipment is in a satisfactory condition.

For visual inspection, we can distinguish between user checks, carried out by clinical staff every time equipment is used, and formal documented visual inspections carried out at prescribed intervals by competent persons. At a formal inspection, the environment should be checked as well as the equipment. Cables, cases and plugs should be inspected for damage and plugs checked internally for sound connections and correct fuse rating.

Electrical testing of portable equipment involves *earth bond continuity* tests and *earth leakage tests* (for class 1 equipment), *insulation resistance* testing and *functional checks*. The permitted earth bond test currents and resistance of the earth circuit on class 1 equipment differ according to the standard used. For PAT, a resistance of less than 0.1 $\Omega$ is specified when tested with a current of 1.5 times the fuse rating, but no greater than 25 A, for a period of between 5 and 20 s. IEC 60601 specifies a pass resistance of 0.2 $\Omega$ with a 50 Hz current of between 10 and 25 A for at least 5 s. These high currents can cause damage to equipment if applied to external points not intended to be protective earths. The newer standard IEC 62353, intended for acceptance and repeat testing, specifies a pass of 0.1 $\Omega$ at acceptance and a pass of 0.3 $\Omega$ at subsequent retests, both with a current of 100 mA. Insulation resistance tests are required for medical and PAT testing standards, for the mains part, to be carried out at a test voltage of 500 Vdc. The medical equipment standards require a minimum resistance of 20 M$\Omega$ under these conditions, whilst the PAT testing standard, whilst allowing a lower resistance limit of 0.3 M$\Omega$ for class 1 heating equipment of less than 3 kW capacity, has a limit of 1 M$\Omega$ for all other class 1 equipment, with class 2 equipment at 2 M$\Omega$ and class 3 equipment at 250 k$\Omega$. The allowed earth leakage currents in PAT testing are geared to domestic and industrial equipment use and are considerably higher than for the medical equipment standard (3.5 mA for some class 1 appliances). Finally, the frequency of testing: PAT standards specify evidence-based testing intervals based on initial risk and type-based assessment that can be modified according to the number of actual faults found on an analysis of test records. In the past, many organisations have settled on a yearly test for medical devices, but this is open to modification on the basis of risk assessments.

## B.8  Electrical Units

The abbreviations used in this appendix for units of electric current are

A     Amperes
mA    milli Amperes (1,000 mA = 1 A)
μA    micro Amperes (1,000 μA = 1 mA)

Those used for potential difference are V-Volts and for electrical power are kilowatts (1 kW = 1,000 watts). For electrical resistance we have used Ω-ohms, kΩ-kiloohms (1 kΩ = 1,000 Ω) and megaohms (1 MΩ = 1,000 kΩ).

## References

1. IEC 60601-1. *Medical Electrical Equipment – Part 1: General Requirements for Basic Safety and Essential Performance*, Ed. 3.1. International Electrotechnical Commission, Geneva, Switzerland, 2012.
2. IEC 62353. *Medical Electrical Equipment – Recurrent Test and Test after Repair of Medical Electrical Equipment*. Ed. 1.0. International Electrotechnical Commission, Geneva, Switzerland, 2007.
3. UK Statutory Instruments 1989, No. 635. *The Electricity at Work Regulations 1989*. http://www.legislation.gov.uk/uksi/1989/635/contents/made (accessed on August 31, 2013).
4. The Institution of Engineering and Technology. *The IET Code of Practice for In-Service Inspection and Testing of Electrical Equipment*, 4th edn. IET Publications, Stevenage, U.K., 2012.

# *Index*